Quantum Trading

Founded in 1807, John Wiley & Sons is the oldest independent publishing company in the United States. With offices in North America, Europe, Australia, and Asia, Wiley is globally committed to developing and marketing print and electronic products and services for our customers' professional and personal knowledge and understanding.

The Wiley Trading series features books by traders who have survived the market's ever-changing temperament and have prospered—some by reinventing systems, others by getting back to basics. Whether a novice trader, professional, or somewhere in-between, these books will provide the advice and strategies needed to prosper today and well into the future.

For a list of available titles, visit our Web site at www.WileyFinance.com.

Quantum Trading

*Using Principles from
W.D. Gann and Modern Physics
to Forecast Financial Markets*

FABIO ORESTE

WILEY

John Wiley & Sons, Inc.

Published by John Wiley & Sons, Inc., Hoboken, New Jersey.
Published simultaneously in Canada.

Limit of Liability/Disclaimer of Warranty: While the publisher and author have used their best efforts in preparing this book, they make no representations or warranties with respect to the accuracy or completeness of the contents of this book and specifically disclaim any implied warranties of merchantability or fitness for a particular purpose. No warranty may be created or extended by sales representatives or written sales materials. The advice and strategies contained herein may not be suitable for your situation. You should consult with a professional where appropriate. Neither the publisher nor author shall be liable for any loss of profit or any other commercial damages, including but not limited to special, incidental, consequential, or other damages.

For general information on our other products and services or for technical support, please contact our Customer Care Department within the United States at (800) 762-2974, outside the United States at (317) 572-3993 or fax (317) 572-4002.

Wiley also publishes its books in a variety of electronic formats. Some content that appears in print may not be available in electronic books. For more information about Wiley products, visit our web site at www.wiley.com.

Library of Congress Cataloging-in-Publication Data:

Oreste, Fabio, 1963–
    Quantum trading : using principles from W.D. Gann and modern physics to forecast the financial markets / Fabio Oreste.
        p. cm. — (Wiley trading series ; 409)
    Includes index.
    ISBN 978-0-470-43512-0 (cloth); ISBN 978-1-118-09352-8 (ebk);
    ISBN 978-1-118-09353-5 (ebk); ISBN 978-1-118-09354-2 (ebk)
 1. Speculation.   2. Stocks.   3. Quantum theory.   4. Physics.   I. Title.
    HG6041.O74 2011
    332.64′5—dc22

                                                                2011007465

10 9 8 7 6 5 4 3 2 1

Contents

Preface

I have written this book for traders, both professionals and beginners, who would like to start a new adventure, a journey into the vast and unexplored land of Quantum Trading. It's a journey that could change your life forever.

This book could also be considered an exploration of phenomenology and quantum philosophy. Why philosophy and not science? Because the quantum viewpoint affects not only particle science but also contemporary psychology and the way we conceive of the relationship between observed objects and the observer. We will also examine how we think of and experience reality.

Great thinkers such as Max Planck, Carlos Castaneda's Don Juan, like Erwin Schrödinger, W. D. Gann, and Albert Einstein have more in common than you might suspect. This book aims to express rocket science concepts in layman's terms and apply them to your daily lives and trading; it's going to be a lot of fun. You will better understand how reality works and achieve amazing results in many areas of your life. Above all, you can make huge amounts of money using these concepts to trade the markets.

We will be applying these ideas to create a Quantum Trading system that provides a high percentage of winning trading signals. In the first few chapters, I describe my proprietary approach to financial trading based on Einstein's theory of relativity and quantum physics, and how to start a very profitable business with little capital, compared to other businesses.

You will be able to set up your own personal "trading firm" without formally incorporating a company. If you follow the rules in this book and always use stop-loss orders, then you will find yourself trading more successfully. Stop-loss orders prevent large, unexpected losses for each trade and allow for the possibility of only relatively small losses, according to your personal risk tolerance.

If you are an absolute beginner and don't know how to calculate the best stop-loss with respect to your trading capital, don't worry. It is explained in detail within later chapters.

This book is also written for professional traders who may have many years of experience, but are looking for something new in the enormous arsenal of technical analysis theories and tools. I have been successfully using the Quantum Trading approach to manage my client's accounts for years, and I believe that other professional traders can use this theory to improve their performance, too.

The techniques presented here are based on a new approach to financial trading that I have developed during the past 15 years, and they are quite different from the classical technical analysis trading tools, even if they share some common ground.

You can apply Quantum Trading techniques to different time frames, such as weekly, daily, 60-minute charts, or 5-minute charts.

This book is not about methods of technical analysis that you are used to seeing. I am not going to speak about moving averages, oscillators, or the hundreds of indicators you already know about. Perhaps you are searching for something different, a more precise instrument that can indicate when the next top or bottom price could form.

Rather, this book is about quantum prices, time levels, and how the theory of relativity and quantum mechanics concepts can practically affect your favorite stock, commodity, or currency price. It is also about trading models that provide precise indications on trend turning points and how to predict the most likely prices and times for the major and intermediate reversal points.

We start our journey by borrowing some ideas from the theory of relativity and its implications in modeling space. We continue by examining electron behavior in the quantum world. Finally, we compare this notion with some of W. D. Gann's 100-year-old concepts about the market that are still relevant today.

We continue our journey by exploring fascinating lands, learning how quantum laws, according to scientist such as A. Goswami, can also affect our daily life. We review how this quantum approach can help us in obtaining a more meaningful life, freedom, and financial independence, and dramatically improve our life and the ability to get what is really important to us.

At the end of this process, you are going to discover many interesting and safe ways to make money in the financial markets using unconventional yet powerful trading techniques such as Quantum Price Lines (QPLs) and Time Algorithms (TAs).

If you continually improve yourself and reach for a higher and higher level of knowledge and awareness, you will exponentially increase your personal power and improve your energy level. You will experience a quantum leap in the quality of your life. Then wealth will come to you like a moth to a flame.

If you are ambitious and feel you have many goals still to reach in your life, then reading *Quantum Trading* can help you to improve your trading and change your life forever.

Fabio Oreste

CHAPTER 1

The Birth of Quantum Trading

How Einstein's Theories and Quantum Particles Affect Your Daily Trading

I t was April 14, 2000, and the market was about to close. I had just placed a selling order on all of the put options I had bought a few days before and cashed in my profit. I stared in disbelief one last time at the prices on my screen. I couldn't believe my eyes! I was excited because the S&P 500 was plummeting, losing more than 180 points from the historical top in a few days, and the put I had bought had earned 128 percent. I had waited a long time, recalculating almost every day the most likely price level for a reversal and waiting for the moment when the S&P 500 price would meet with the maximum curvature point of the P-Space, the new trading tool I had been developing during the past years. From another point of view, this would have been the point at which the quantum entanglement would be relevant.

Quantum entanglement is a property of the quantum mechanical state of a system that contains two or more objects. Specifically, the objects that make up the system are connected in a way that cannot adequately describe the quantum state of a constituent of the system without full mention of its counterparts, even if the individual objects are spatially separated.

I had been waiting for this quantum correlation between P-space and the S&P 500 price for a long time. In those days I had not yet developed the software that now allows me to visualize in a few seconds the points most likely for a reversal on the chart. To calculate everything by hand, I needed a lot of time.

The week before I had warned my clients to close all of their long positions on stocks and take profits. Most of them were astonished and asked

1

me, "Why do I have to sell when the stock exchange continues to rise and I'm making money?"

"Stock prices are likely to drop significantly soon," I explained, "and before we see such price levels again, many years will pass." At that time they surely thought I was crazy, but in hindsight I was proved correct. If you study my Quantum Trading techniques you will discover why.

I wanted to go out and get some fresh air, despite the fact that it was raining, when the phone rang.

"Hey, Fabio, have you seen the S&P 500? It seems you were right. It reached the levels you forecasted. Try not to disappear tonight because we need to celebrate! I saw how much the puts you bought for my managed account are worth now, and we made a fortune. All of us will come over to your house around eight-thirty tonight. We'll bring the champagne and dessert, and you'll cook the risotto. We need to talk. You have to explain to me how your theory works again. By the way, does your theory also work for currencies?"

"Yes, it also applies to currencies, and to everything else traded on the stock exchange. Your plan sounds great. See you at eight-thirty," I responded to Dave, a friend of many years and also one of my best clients. I left my house and walked briskly through the pouring rain. I needed to decompress after such an intense trading day. The raindrops beat down sharply on my head, helping me to reflect.

Take a look at Figure 1.1. On April 10, 2000, I bought put options on the S&P 500 expecting a big drop for stocks and the entire index because

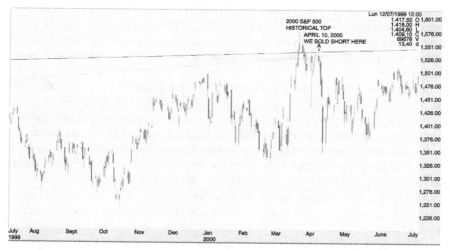

FIGURE 1.1   S&P 500 Future: Major Top Forecasted in 2000 Using QPLs

two major Quantum Price Lines (QPLs) were simultaneously touched by the index price at point A on the same day.

This is a very rare event and when it happens the market is ready for a big movement within a few days. Instead of the put option we could have sold short a future contract on an S&P 500 to take advantage of the significant drop following the contact of the price and the two QPLs. The two QPLs indicate the point of maximum curvature of the S&P500 P-space which, when reached by the price at point A, indicates a strong reversal pattern. How did we calculate them? We discover the answer in the next few chapters.

Walking down the street, I was happy because I had finally proved to myself that my trading theories, based on Einstein's space-time discoveries and on the behavior of an electron leaping from one energy level to another, were working well and were profitable. After years of study and observation I had finally developed two complementary trading models that worked very well together. Physicists were still split on the supremacy of quantum mechanics over Einstein's theory of relativity, and many contradictions were still in place in the standard model of quantum physics because the Higgs boson has not yet been found. Even though Einstein didn't believe in quantum physics, my trading model uses both Einstein's theory and quantum mechanical approaches, successfully harmonizing both ideas for trading purposes.

The models that I had elaborated for trading were only partly drawn from physics. I had developed these trading models in an autonomous way beginning with the assumption that the entire universe is connected by a gigantic entanglement—or interconnection—governing not only the subatomic particles behavior, but also other complex and immaterial relationships, such as financial transactions.

I was also inspired by W. D. Gann's statement about the close relationship between atoms, electrons, and stock price behavior. Gann was one of the most legendary traders in the history of Wall Street. My way to calculate QPLs is partially based on Gann's confidential work, although the concept of P-space, based on the equivalence between Einstein's light deflection and price deflection phenomena, is completely new and unknown by previous traders and authors, including W. D. Gann. Furthermore, I rigorously propose a quantum scale for QPL calculation, based on Leibniz's original chart of 64 codes on which binary code is based, refusing to use a linear approach, and this is another new concept in trading. Two to the power of n is the number of ways that the bits in a binary integer of length n can be arranged. We use them to calculate our QPL's price orbital, the same way they are used to measure computer memory, processor power, and computer disk drives.

The trading of stocks, bonds, currencies, and commodities takes place in our everyday reality, a relatively reliable world that apparently doesn't seem affected by the paradoxical laws postulated by Albert Einstein's relativity theory and Niels Bohr, Max Planck, and Erwin Schrödinger's quantum mechanics. What happens in our daily life can be elegantly explained by classical physics, which is full of reassuring certainties about linear models ruling space, time, and other things, such as locality, on which we base our representation of three-dimensional reality. For example, it is very easy for classical physics to measure the mass, strength, and velocity of an object, such as a bullet, and precisely predict its trajectory through space. Unfortunately, these certainties cannot explain nonlinear models like the ones ruling stocks, commodities, Forex (FX), and financial markets in general.

According to what I had discovered, the prices of stocks, or anything else traded on the different exchanges, were not only influenced by news about profits, GDP, the Fed's minutes, mergers and acquisitions (M&A), write-downs, interest rates, inflation, consumer confidence, or other fundamental data, but also by nonlinear entanglements based on the theory of relativity and quantum physics. Using trading models based on these theories, we can understand the financial securities price behavior. My trading models are based on the entanglement formed between different categories that operate in a multidimensional, mathematical space that is curved because of the presence of objects with a specific mass, which I call P-Space.

It seems that the same day unexpected news affecting the financial market is released, provoking a top or a bottom, my Quantum Trading models show that the price has reached the maximum curvature point of specific space-time in which the price of a financial security moves, that is, the P-Space.

I used two different sets of concepts to develop my Quantum Trading view: one from Einstein's Theory of Relativity and the other from Quantum Mechanics. Apparently these two visions of the universe are incompatible, but my trading based on these concepts worked very well and I was making money.

On the other hand, by studying Nobel Prize–winner David Bohm's theories, one could find some clues to harmonize these two irreconcilable views of the universe.

I had left behind even dear old technical analysis, with its arsenal of oscillators and indicators, which was unable to forecast the final top of a big movement, or the major bottom in the markets, despite many attempts.

It's not as if technical analysis was not interesting for me. In fact I had studied it passionately for years. Yet I was looking for a "Theory of Everything" to explain the financial market's movements using consistent and elegant models. For instance, technical analysis wasn't able to explain why a double bottom sometimes is a very strong support and the price bounces, but other times it is easily broken and price collapses.

Instead, the models I had developed, inspired by physics, were able to forecast if a major or intermediate top or bottom would be likely to occur at a certain time using the quantum level of the orbitals of price—very similar to an atomic orbital—and calculating the points of maximum curvature.

Correctly forecasting the time and price of the turning point of a stock index is a dream for every trader because it means you could make a fortune if you know where the bottom or the top is located.

The best part of this approach is that Quantum Trading models can calculate these turning points weeks or even months in advance with high probabilities of success. You just need to wait until the price reaches the QPL you have already drawn on the chart far in advance. When the QPL price level is reached, you are likely to see a significant reversal, as shown in Figure 1.2.

You could have sold short HP stock on the QPL resistance at point A and closed the long position on the QPL support. HP stock wasn't able to break the QPL resistance level.

This means that if you had sold HP short on the QPL resistance you would have made money because the price dropped. If the price would have broken the QPL resistance then it would have continued to rise until the next available QPL. By putting a stop-loss order one dollar higher than the QPL resistant price you would have exited the short position and you

FIGURE 1.2 Hewlett Packard Stock: 2010 Top

could open a long position at the stop-loss level, buying double the amount of stock you previously sold short. This is called "stop and reverse" and all experienced traders are very familiar with it.

If you wonder if I have really found the Holy Grail of trading, I have to clarify a very important point. In quantum physics, we speak in terms of probabilities that the electron can be found at a certain position using the probability wave function. In the same way, in my Quantum Trading models I speak of the high probability that a significant reversal can occur, or a minor probability that pushes the price toward a higher orbital level, exactly like an electron's quantum leap.

In physics an atomic orbital is a mathematical function that describes the wavelike behavior of either one electron or a pair of electrons in an atom. Quantum physicists use this function to calculate the probability of spotting any electron of an atom in any specific place around the atom's nucleus. These functions can form a three-dimensional graph of the likely location of an electron. Specifically, atomic orbitals are the possible quantum states of a single electron in the collection of electrons around a single atom.

In the same way, we use QPLs as if they were atomic orbitals to understand the behavior of financial securities prices.

If they tell you that you can use all of these physics concepts to forecast the next top or bottom of your favorite stocks, you would probably think it is just a fairy tale, but when using the Quantum Trading models for a significant period of time, the results will probably help change your mind.

I am not asking you to believe me up front, but rather to follow me chapter by chapter and then experience the results for yourself.

A LESSON IN ELEMENTARY PHYSICS

Around 2,500 years ago in India, Buddha said, "Do not believe unless your experience can prove it," a beautiful example of structural skepticism that should be applied to everyday life.

If your knowledge of physics is rudimentary, relax. I will use simple terminology in the pages that follow. I will attempt to communicate all of the concepts that usually require complex equations in a simple way without using "rocket science" notation.

I have included some very simple mathematical equations in order to meet formal criteria requirements. At most, I will sprinkle some simple formulas here and there, but nothing complicated.

All the charts here could be visualized in a three-dimensional way, but this surpasses the limits offered by a book.

Many years ago, I did not speak to anyone about my discoveries other than a few friends with whom I had shared travels, studies, and adventures. Certainly colleagues in my field would have thought I was crazy or doubted my masters degree in Business and Finance, which I earned from Luiss University in Rome.

Moreover, at that time I was missing some elements that would have made my model more elegant, ridding it of some contradictions that I was not yet able to solve, and that I came across only a few years later, thanks to string theory.

Nevertheless, my short S&P 500 trade mentioned earlier in this chapter and based on this theory was placed very close to the historical top of 2000 and was a powerful reversal point for the index before a big plunge.

The high profits I earned from my S&P 500 trade in only a few days had just materialized into my account. This, more than anything, boosted my confidence in my Quantum Trading models.

Absorbed by these thoughts, I arrived at the store; I had to shop for ingredients before my friends came to dinner in a few hours as well as pick up Monika, my girlfriend, who had just finished a photo shoot in the Caribbean. Everyone expected risotto for dinner and I was more than happy to cook for them.

After my friends arrived I occupied myself with preparing the risotto, but I did not use the *risotto al Barbaresco* recipe I intended to use. Dave had been sent truffles from Italy and brought them to my house so that I could put them on the top of the risotto instead of using Barbaresco wine to cook and aromatize the rice. My friend considered me a very demanding gourmet and the risotto I cooked that time was very satisfying because the ingredients were really terrific. The truffles David brought were picked only the day before and were very fragrant and tasty.

Dave is a really great guy! Smart and interesting, he was at the time a very successful real estate entrepreneur and a good friend who traveled to Europe and Asia with me. In Europe we liked to search for the finest dishes and best wines. We would visit many different vineyards and restaurants in our quest to find the best. In Asia we traveled together through the most stimulating places in Tibet, India, Nepal, and China to study ancient Eastern philosophies and the science of the mind.

Einstein, Bohr, and Planck were all interested in studying Taoism and the Abhidharma and Veda philosophy and cosmology. These ancient philosophies contain many interesting concepts that relate to the subatomic world, and the wisdom contained in these systems seems to be the precursor of modern science and physics.

David Bohm, a Nobel Prize–winning scientist, spent many years formulating a higher order of physics and was close to discovering a solution that would make the theory of relativity and quantum mechanics compatible.

For many years he studied Vedanta, one of the most important philosophical schools of ancient India.

Dave offered to go with my butler to the cellar and personally choose the vintage of Barbaresco to accompany the meal. When he returned I was battered with questions.

"Now, explain to me, Fabio, what Einstein's studies of space-time and quantum physics really have to do with the stock exchange?" asked Dave, happy that two more bottles of wine were on the table. "I knew you were crazy, but fortunately I made the same trade you placed with my managed account with my personal account and it's made me a small fortune. I'm starting to think that you got your hands on something really big considering how much we gained from the last trade made with your precise forecasts from three weeks ago." Dave liked to occasionally trade Forex and S&P 500 futures on his own, using some basic technical analysis tools in addition to the account I managed for him.

"Well, it's a long story and I don't want to bore the others.... We'll talk another time," I responded, knowing Dave would not give up so easily.

"No, no. We absolutely have to know how Quantum Trading works," Dave insisted.

"Now we're curious and you have to explain everything," added Elena, my best friend. She was usually very interested in my travels and studies. "So let's eat the Sacher cake I baked today, open a bottle of champagne, and make a toast to your trading system," concluded Elena, smiling happily.

We quickly agreed to Elena's suggestion. The Sacher cake was exquisite, especially since it is difficult to find true Sacher cakes outside of Austria. Fortunately Elena's grandma was from Vienna and was an amazing cook.

"Okay, okay! We'll start at the beginning and move step by step," I conceded to my friends. In reality, I was happy to talk about my trading models. We toasted, and while I was appreciating the apricot marmalade layer of the Sacher cake, I thought that celebrating with old friends that evening at home had made it all worth it.

"What I have discovered is that the price of a stock or a security can be seen either as a light particle, called a photon by scientists, or as an electron. First, Newton's classic physics, Riemann's curvature tensor, and Einstein's theories of relativity leading us to the concept of the curvature of space help us to understand some aspects of a security's price behavior. Second, quantum physics allow us to better understand the characteristics of stocks, commodities, and currencies price behaviors.

"If we want to understand how Quantum Trading works, we need to refresh our memories of some of the basic concepts developed by the founders of classic and quantum physics.

"Everything actually begins in England with Isaac Newton, the father of classical physics and one of the most influential people in human history. He was an absolute genius who worked on his theories for years, adding to the scientific knowledge of his time by exploring the laws that govern both our everyday reality as well as our solar system. He had explored all of the sciences and considered mathematics insufficient to explain his findings. As a result he and Gottfried Leibniz developed a new type of mathematics: differential calculus.

"Newton was not known for many years by his contemporaries because he preferred to devote himself to research at the expense of his social life. It was only in 1687 after publishing his masterpiece, *Philosophiae Naturalis Principia Mathematica* (Mathematical Principles of Natural Philosophy), that he gained success and fame all over Europe. For the next two centuries this book would be considered the most important scientific treatise ever written.

"Following the publication of his treatise, he became a real celebrity. Two years later he was elected to Parliament, and in 1703 he became president of the Royal Society. In 1705 Queen Anna knighted him as Sir Isaac Newton, the first scientist to receive that honor."

"Ah, okay, now I remember," said a tipsy Barbara, Dave's long-time girlfriend. "Newton was the one who shot the apple that fell on his son's head with a bow and arrow."

"No, Barbara, you've mixed up William Tell, the Swiss hero, with Newton, the scientist," explained Dave.

"Well, actually, she's not far off because Newton was inspired to formulate his theory of gravity by observing the fall of an apple from a tree," I added, laughing.

"But didn't he also study alchemy? It seems that following his death they found numerous writings and research on this topic. Is it true?" asked Elena.

"Yes, that's true. Imagine that even though he is universally considered the real father of modern science, he wrote more pages on alchemy, occult sciences, and theology than physics or mathematical subjects. But that's another story that will take us too far off topic, so let's get back to classical physics."

"His first big discovery was the theory of Universal Gravitation and the movement of planets. According to Newton, all celestial bodies are attracted to each other. The source of gravity instantly passes from one point to another of the universe and keeps the planets in our solar system fixed in their orbits while preventing us from floating away. His idea of the universe is as a perfect mathematical place, very similar to a giant cosmic clock, where every mechanism unwinds in a precise and predictable way. It's a mechanical universe based on the principle of cause and effect where it is possible to represent and measure exactly the movement of objects. It's a universe ruled by an absolute deterministic order in which paradox, contradiction, and indetermination are unknown."

"What do you mean by absolute order?" asked Monika.

"Newton, to instill order on chaos, proposed a concept of absolute space and time. For this reason everything was simple: It was possible to unequivocally identify motion and time because absolute space always remained the same: immobile and without a need to relate or refer itself to any other external object fixed of moving. Absolute time works in the same way as absolute space."

"Well, why is it not always so?" interrupted Monika. "If I take a flight from New York to San Francisco and it takes five hours from take-off to landing, aren't these five hours the same for everyone? The clock ticks and time passes in the same way for those traveling in the air as it does for those on the ground."

"This is one of the crucial points in the theory of relativity. For Einstein, time is not absolute and does not tick in the same way for everyone. After Newton it took three centuries before Albert Einstein proved that space is not a three-dimensional entity separate from time. Space and time form a continuum. The closer an object's movement is to the speed of light, the more time slows down for them. Do you recall the paradox of the Einstein twins?"

"I think I know the story," Elena quickly responded. "The first twin remains on Earth while the second travels in a spaceship that moves at the speed of light. After many terrestrial years, the second twin returns to Earth and finds his brother crooked and aged with gray hair, while he is still young. It's a paradox because the twins should be the same age, given that they were born minutes apart."

"Yes, it's just like that. The first twin remained on the earth and for him time passed 'normally,' while for the second one in the spaceship, time passed very slowly and almost stopped compared to time on Earth, because he was traveling at the speed of light," added Dave.

"In fact in 1911 Einstein asserted that 'if a living organism, after an arbitrarily long flight at a speed approximately equal to the speed of light, could return to his place of origin, he would only be slightly altered while his corresponding remaining organisms would have already given birth to new generations.'

"The point is that even if the twins were born more or less at the same time and were the same age, for the one on the spaceship time passed slower in comparison to the one on Earth. Einstein concludes that it is important to evaluate the passage of time of an object on the basis of its speed with respect to that of another observer. In this way, the concept of absolute time was destroyed and the "relative space-time" concept remained.

"I used the idea of 'relative space-time' as the base from which I developed my P-Space theory. I use it to calculate the most probable time and price for a reversal in the financial markets," I explained.

"Among the various consequences of this revolutionary hypothesis is that the passing of time varied according to the state of motion—or state of rest—of the observer, depending on the velocity with which the latter moved.

"It's exactly to explain that point that Einstein suggested the famous "Twins Paradox" we have just spoken about, even if it's not a true paradox, since it is completely explained in the context of the two postulates of the theory of special relativity. There are two twins, initially in the same place and with two identical clocks that are synchronized. One of the two twins remains on Earth, while the other leaves for an interstellar journey on board a spaceship, whose elevated velocity reaches 80 percent that of light. On his return to Earth, his clock indicates that 30 years (of "real" time) have passed since his departure, while the clock of his twin, left on Earth, indicates 50 years have passed since the departure of the spaceship.

"Since the astronaut twin does not perform a uniform motion, but has to accelerate or decelerate to carry out the departure and return, the situation is no longer symmetrical: the astronaut will have, in effect, lived less than the twin brother left on Earth.

"But why do you need to know all this stuff if you want to make money trading the financial market with your system?" asked Barbara.

"If you begin to see things in this way and try to apply these concepts to stock, commodity, and FX trading, you can revolutionize your way of interpreting financial phenomena and start seeing them in a different light."

"So all of this plays a role in your trading system and the money you both earned, with which my dear Dave will buy me that wonderful Bulgari diamond ring that I saw last week?" exclaimed Barbara.

Dave coughed and his face immediately turned red. He knew that taking Barbara to the Bulgari shop would cost him a fortune. In that moment it seemed he regretted telling Barbara how much he had earned that week with the help of my algorithms.

"Barbara is right," I said, enjoying her dazzling smile.

Dave quickly snapped back, saying that given how I was agreeing with her, I could be the one to accompany her to the Bulgari shop and use my credit card instead of his.

"No, Dave," I said. "I was referring to the fact that Barbara was right in saying that, until now, I have only spoken about relative time and I've not yet arrived at the crucial point.

"The point is that according to Einstein, even space is not uniform, the opposite of what Newton assumed."

"In fact, in the general theory of relativity, Einstein specifies that mass is a form of energy and that the force of gravity, due to the presence of mass in space, has the capacity to curve space. The implications of this concept are numerous, but it is really this property of the curvature of space that led me to the idea of creating a virtual mathematical space in which the price of a share moves similarly to a particle of light, a photon. When the price moves and eventually reaches a point of curvature, it deviates from its trend and inverts, creating a top and a bottom."

"Explain yourself better," said Dave, "I studied technical analysis a little bit, but I'm unable to see any resemblance. What is the relationship between the curvature of your mathematical, virtual space in which the price of stock moves and marks a major reversal of the trend?"

"Bernhard Riemann's approach to topography is very enlightening and can help us understand the entire process. Applying Riemann's idea of the curvature tensor to the space where the price of a stock moves helps us reach a deeper understanding of my trading theory. Try to picture the price of a stock as a ball in constant motion on a rugged terrain full of *cavities* and *bumps*. At times the ball ascends a bump, which corresponds to an uptrend. When the ball reaches the peak of the bump it corresponds to the highest price of the stock, or the top in its chart. Once it arrives at the peak it begins to descend, sliding down toward the valley, which corresponds to a downtrend. Then you may see it roll on the flat plain, and this corresponds to a lateral trend. When it descends to the lowest point of the cavity, this corresponds to a bottom in its chart. Still in motion, the ball returns and ascends. It's a matter of topography.

"The ball is the price of a stock that rises and falls. Do you understand now?"

"Yes, now it's a bit clearer. But didn't you say that in your model the price was like a particle of light?"

"Einstein probably made a similar reasoning to figure out the phenomenon of the deflection of light in the presence of masses that curve space. In fact, he used Riemann's geometry and curvature tensor, the basis of modern topography, to finally express his theory through a definitive formal model.

Despite the fact that he had already discovered the core of his theory on light, space, and time a few years previously, Einstein could not progress with his theories for years until Marcel Grossmann suggested to him to consider the revolutionary Riemann studies and his curvature tensor. Einstein didn't know Riemann's geometry until that moment. Grossmann suggested that Einstein would need a space-time model possessing not only the flat, Euclidean properties of special relativity, but a space-time possessing non-Euclidian properties, like Riemann's geometry.

One of space-time's main features is that, while it appears curved on a grand scale, it appears flat on smaller scales. That is exactly what happens if someone stands on a football field and looks around: the Earth will appear flat. The first consequence is that for the description of events confined to local regions of space-time, special relativity remains valid. But things appear differently for large regions over which the curvature of space-time becomes significant and visible. In the same way the football field looks flat to a football player, but America looks curved if observed by an astronaut. So, it's easy to understand that the larger the radius of a sphere, the smaller its curvature. In the same way, the larger the radius of a sphere, the greater the area surrounding any point that appears to be flat, if observed locally.

"Einstein, in his book *Relativity*, came to a very important conclusion: The distribution of matter in the universe determines the amount to which space-time is curved: The greater the density of matter in a region, the higher the curvature of space-time. Thus space-time is distorted more around the Sun than the Earth because the Sun has the larger mass. This means that gravity no longer exists as such; it is transformed into the curvature of space-time."

"Can you please clarify what you said about the deflection of light, given that it's so crucial to your theory?" asked Elena.

"Of course," I replied. "Let's imagine Newtonian space for a moment. It's uniform and, for the sake of explanation, we will represent it with two dimensions (see Figure 1.3), even though it always exists in three dimensions. We can compare this space with a tablecloth that forms a flat plain. Let's hold the tablecloth in the air and take a steel ball that represents the sun and place it in the middle of this space (see Figure 1.4).

"According to Einstein, due to heavy mass, space is distorted just like our stretched-out tablecloth is distorted when it is indented by the steel ball. The end result is that space, in our case the tablecloth, now curves" (refer back to Figure 1.4).

"This leads us to two considerations. Regarding gravity, the first effect of the curvature of space is that it supplies us with the line of minimum resistance on which the planets move around the sun. Regarding the movement of light, which is crucial for understanding our trading model, particles of light, originating from a distant source and passing close to the sun, deviate from their rectilinear path. This happens because the gravity of the mass of the sun curves the space in which the particles of light travel. In this way the particles of light are deflected—and the price of a stock follows a similar pattern" (see Figure 1.5).

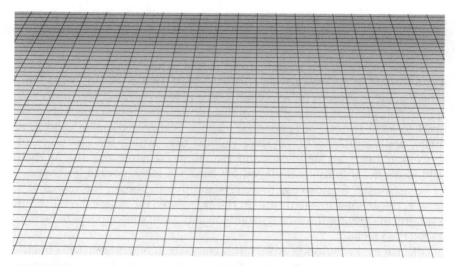

FIGURE 1.3 A Schematic Representation of Flat Space

"You mean to say that the price of a share behaves like a particle of light and it's enough simply to apply the formulas of relativity to forecast future stock and currency prices?" asked Dave.

"No, wait a minute! My model only uses some concepts inspired by Einstein's work on relativity. A few concepts were taken from quantum physics regarding the behavior of particles, and the rest stemmed from

FIGURE 1.4 The Sun Curves the Space around It

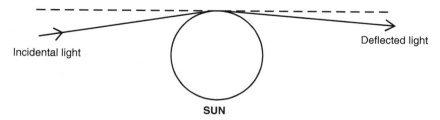

FIGURE 1.5 A Similar Pattern Unfolds with the Price of a Stock

other diverse disciplines. Even though the concepts I use are inspired by relativity and quantum physics, the equations I use are not necessarily related to them, but rather use a simplified mathematical operation. I'm not pretending to have discovered what Einstein never said about the stocks. That would be absurd!

"My model was inspired by the behavior of a particle of light, whose trajectory deviates because of the presence of a celestial object with a mass that curves the space around it. The model only serves to describe a behavior and provide algorithms that, if applied to the P-Space, a virtual space-time, enable us to calculate the most likely points for a reversal on the chart of a financial security. The price, after touching certain levels, reverses itself because it's as if it intersected a curvature that has deformed the space" (see Figure 1.6).

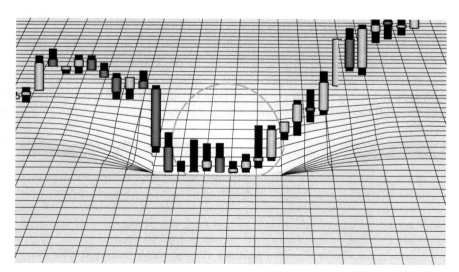

FIGURE 1.6 A Curvature That Has Deformed the Space

"Anyway, the Einsteinian concept of light deflection plays a crucial role in my trading system. It inspired me to create the structure of P-Space where price moves, ahead of time. It also inspired me to calculate the point of curvature of P-Space using mathematical operators stemming from physics. When the price reaches the point of curvature, the trend can reverse.

The concept I borrowed from topography relates to the example of the ball, representing price, which rises and falls according to its journey through the uneven terrain. This example is only useful as a description and does not allow me to make forecasts unless I measure it first. It allows me only to observe a bull or bear trend as it happens just by looking at the P-Space chart."

"Do you mean to say that the price level at which the inversion will most likely take place is constant and that sooner or later it will invert its course?" asked Dave.

"No, just the opposite. The price level at which the curvature begins is not constant and varies with the passing of time. It can be calculated using my curvature equation of the P-Space. The curvature deflects the price and causes in the P-Space a reversal. We perceive this reversal in our common stock charts as a top or a bottom" (refer back to Figure 1.6).

P-SPACE: A QUANTUM TRADING TOOL

"You keep mentioning this P-Space. What do you mean when you say P-Space?" asked an ever-more intrigued Dave.

"It's a multidimensional, virtual space composed of securities prices, time, and objects in movement, which curve the space due to their mass and gravitational effect. You can apply similar laws to P-Space, as those indicated by the theory of relativity and the laws ruling electron movement, though not the exact same equations."

P-SPACE

P-Space is a matrix that can be linked to the chart of any security. On this two-dimensional chart, we can project lines that represent the points of curvature for every time frame (minutes, hours, days, or weeks). Under certain circumstances curvature points can transform into important levels of support or resistance, which are invisible using a traditional, technical analysis approach.

The support and resistance levels are spotted using Quantum Price Lines (QPLs; see Figures 1.1 and 1.2).

"So if you say that this P-Space is virtual, that means it doesn't actually exist and you invented it!" observed Barbara, perplexed.

"Well, sure it doesn't exist in the same way that your earrings and rings exist, Barbara, but it exists just like mathematical matrices and virtual reality exists.

"P-Space is an interactive, virtual structure ruled by entanglement, or nonlocality. It is composed of different pieces of the reality that surrounds us, such as objects in the solar system that move according to the mathematical laws discovered by Kepler and perfected by Newton, time, and the prices of a stock or currency whose movements are similar to that of a particle of light.

"Having reconstructed our universe as a virtual reality, we can start to see how a price will behave if it intersects a large gravitational mass. We will see the effects of its deflection from its earlier path by applying principles discovered by Einstein. Finally, we can verify the results by observing the actual movements of the stock price and how they relate to the generated QPLs (see Figures 1.1 and 1.2).

"With accurate observation you can see that the effects of deflection in P-Space are amplified with respect to our solar system. But P-Space is a virtual space, a mathematical laboratory created ad hoc to measure the interactions of space-time with stock prices using simple equations that are exceedingly accurate. The function of the P-Space is to help us forecast the major and intermediate reversals of the financial markets."

"And so how does quantum mechanics fit into this?" asked Dave.

"Well, we're not there yet. But so as not to leave you hanging, you can consider the price as an electron that jumps from one quantum orbit to another. This exercise allows us to calculate the probability of a stock price jumping from one QPL to the next one. We can do that by applying the conceptual arsenal of quantum physics, such as Bohr's atom model, Plank's constant (denoted h), and Schrödinger's wave.

"The Planck's constant (h) idea led me to understand that a stock's price, like an electron, moves from one energy price-level to another, thanks to the fact that it either gains or loses a discrete 'quanta' of energy. In P-Space the different energy levels can be measureable through QPLS that provide powerful support and resistance price-levels."

"Does 'string theory' play any role in your models?" asked Dave.

"Actually, the QPLs we can draw on the chart of a financial security can be seen as the result of strings loaded with information that illustrate the behavior of the price according to classical physics, the theory of relativity, and quantum mechanics, at the same time."

"Do you think that your theoretical analysis system can consistently make money on the financial markets or can it only be sporadically applied?"

"It's a tried and true trading system based on the concepts of reversal and acceleration (continuation) of the trend."

"So how does your trading system actually work?"

"Just like Schrödinger's quantum cat!

"My Quantum Trading system is based on a model similar to the wave function that we apply in the proximity of the P-Space curvature points. Even though it might seem quite strange for a theoretical physicist to combine Schrödinger's wave function and Einstein's space-time curvature, we do it, and it works very well in our trading model. Fortunately, we are traders and we have more room than a scientist to arrange our models to make them effective to make money in the financial markets. We cannot be 100 percent positive if a top will form or not in advance, just as one does not know if Schrödinger's quantum cat is alive or dead.

"Many physicists reacted to the quantum cat paradox with irritation because they believe that it does not have any 'real' consequences on quantum mechanics. Stephen Hawking said: 'When I hear of Schrödinger's cat, I reach for my gun.'

"Schrödinger designed a mental experiment using a cat as an illustration. A cat is closed up inside a box containing a sample of some radioactive material and a tube containing deadly hydrogen cyanide. The process of radioactive decay is itself quantum mechanical and accordingly can only be predicted to occur in a probabilistic sense. When an atom within the radioactive sample decays, a signal causes a hammer inside the box to drop on the tube, releasing the toxic gas and killing the cat. According to the layman, the cat is either dead or alive, but according to the principles of quantum theory, the whole system comprising the box, the cat, and its other contents can be described by a wave function. Assuming that the cat can only exist in two quantum states—alive or dead—the wave function for the box system involves a combination of these two possible and mutually exclusive solutions arising from observation. The cat is both alive and dead at the same time, a strange and irrational combination of these two states. Just as the electron is neither a wave nor a particle until a measurement is made, in the same way our cat is neither alive nor dead until you open the box and look.

"According to quantum physics, the cat is in an indeterminate state, alive and dead at the same time, until we open the box. In our real trading activity we can see if after touching the QPL the price forms bars that confirm the reversal itself according to traditional price dynamics. In this case we can use several filters to decide if we should open the trade after the contact between price and a QPL. In case of a break of the QPL we just follow the trend instead of trading for a reversal, and our target will be indicated by the next QPL, exactly like an electron jumping from one energy level to another one, according to Planck's constant (h).

"If you are a more aggressive trader, you can open a short position exactly at the price where a QPL passes, always using a stop loss, and assume that the trend will reverse when the price touches the QPL level. I only take this position, however, if other algorithms of time and price agree on the same information. I have to use several equations at the same time. Luckily, the software we now use allows us to visualize all this in a matter of seconds.

"When the price is in proximity to the curvature point, we limit ourselves to observing its behavior at that specific point. About 70 percent of the time, it will invert, and the rest of the time it will break toward higher or lower levels. As in quantum physics, it is a problem of the cloud of probability. Our model is inspired by the wave function that describes price probabilistic behaviors."

Trading with QPLs

When trading day to day, we usually wait for the time when the price reaches a QPL, and at that point we observe the price behavior. For instance, if the price touches a support QPL and is unable to break it within the first 60-minute bar, then, at the beginning of the second hourly bar, we buy long. If it happens that you are a more aggressive trader, you can open a long position just as the price touches the QPL, betting on a reversal of the trend.

We sell short if we're coming out of an uptrend. We buy long if we're coming out of a down trend precisely at the price that corresponds to the point of maximum curvature.

We can spot these points utilizing QPLs. We always utilize stop-loss to close the position in case the trend contradicts our initial position, and we can also use a stop-and-reverse order. If short-selling on stocks were disallowed, you could still buy put options to open a short position, or you could sell short the entire stock index by selling a future.

You can better understand the preceding discussion by studying the QPLs drawn on a CME Group EUR-USD future chart, as displayed in Figure 1.7. At point A we came from a bull trend showing higher highs and the euro breaks, without hesitation, the QPL at 1.5073. The run of the euro against the U.S. dollar continues until 1.5984, where another QPL blocks the price surge offering a strong resistance at point B. The euro is unable to break it and so price drops. At point C the euro tries its last attack on the same QPL, but fails to break it again and is finally ready for a reversal. At last, the price collapses.

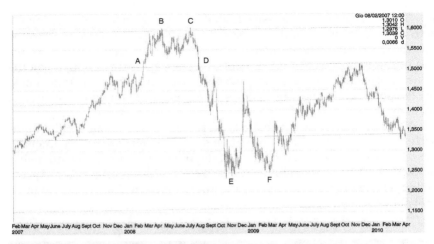

FIGURE 1.7 CME Group EUR-USD Future Daily Chart and QPLs

Isn't it amazing that a QPL passes exactly at the level of the all-time EUR-USD top before the double top was formed? And is it not astounding that another QPL offers support at points E and F at 1.2471?

At point F the trend reverses, reaching the next major top at 1.5147, where, at point G, just "by chance" another QPL is there to offer very strong resistance. Then the euro collapses again.

The price jumps up and down on the QPLs like a photon's trajectory deviated by a curved space, or like an electron jumping from one atomic orbital to another.

This is the magic of Quantum Price Lines. They will enchant you as you calculate them and draw them in a chart. Continue reading the next chapters and you will be able to do it, too.

"But doesn't this combination between physics, the theory of relativity, and the stock exchange seem strained?" asked Elena. "The first two are made up of mathematical equations and rational logic, and the third of volatility, unpredictability, fear, enthusiasm, and investors' euphoria, which push prices up and down in response to the latest news and data that surface on the market."

"Elena, what you say about news driving the markets is correct, but my model works all the same. The former doesn't exclude the latter. Rather, it seems that the phenomena happen simultaneously, not because they are connected, but because they are coemergent, following the principle of Carl Gustav Jung's notion of *synchronicity*.

"If you think that all of this is bizarre, I almost agree with you. But if you looked closely at recent financial trading history, you would discover some unsettling things. For example, did you know that one of the first formulas for calculating the pricing of an option, put together by Black and Scholes, is based on Brownian motion? Brownian motion is a mathematical model used to describe the behavior of single, heavy particles present in fluids or fluid suspensions—for instance, the casual movement of pollen in water. The phenomenon was studied by Louis Jean-Baptiste Alphonse Bachelier and then by Einstein, who in 1905 wrote a study titled 'Investigations on the Theory of the Brownian Movement.'

"The Brownian movement model of prices and financial stock is an essential element in current derivative products pricing as well as in other general financial activity. It is a movement adopted on probabilistic calculation. The mathematics behind the Brownian movement used in the financial field differ from the ones commonly used in the physics field and are based on the stochastic calculation of Rusian L. Stratonovich; in finance, Black and Scholes would use it for their stochastic calculation based on Ito's equations. Their initial intent was to check if the options and the warrants issued on various stocks offered a possibility for arbitrage. In many cases at that time, the windows of arbitrage were much bigger and more frequent compared to today, especially on warrants, and they made a fortune.

"To summarize, I invite you to consider Mr. Black and Mr. Scholes, no less bizarre than me. They used the Brownian model governing the movement of gas particles to calculate the most likely price of a stock option, while I use the theory of relativity and some concepts taken from quantum physics to calculate powerful and lucrative entry points for stocks, currencies, or whatever is traded in the financial markets with significant volume.

"My model, based in P-Space, is able to calculate the prices and times most likely to form a major or intermediate reversal in various financial markets, and it even applies to your favorite stock or currencies."

"You do realize that being able to accurately forecast reversal points of the trend means making a lot of money. . . ," observed Dave.

"Yes. And we have barely started to approach trading in a new way. Cheers to Quantum Trading!"

"Didn't Einstein state that quantum physics was real, thus the world was crazy?" asked Dave.

"Yes, it's true. But Feynman, one of the most important scholars of quantum physics, noted that even if a few people could understand the theory of relativity, there wasn't anyone who could fully understand how quantum mechanics really worked. Years later, things don't seem to have changed: Physicists apply the calculations established by the "founding fathers" of quantum mechanics, but they don't ask themselves how and why these procedures are able to operate or what they really mean. This leads to the application of the so-called 'standard model.' Questioning the nature of the calculations at the base of his model was actively discouraged by Bohr who advised people to limit themselves to the facts without getting lost in superstructures. These superstructures were instead much more important to David Bohm, who wasn't satisfied with the standard model of quantum physics and devoted his entire life to discover and perfect an alternative quantum model able to answer more questions. This caused him to be ostracized by the scientific community despite his brilliant research on plasma and a Nobel Prize already won."

I looked at Monika, who winked and smiled back. Quickly, I took her cue and wrapped up our gathering.

"All right, everyone, it's time to say goodbye. Enough talking about physics. It's time to dedicate ourselves to quantum biology!"

CHAPTER 2

How to Psychologically Prepare for Successful Trading

The most recent research in neuroscience shows that our brain seems to work in accordance to quantum laws. It means that all of us build up, day after day, our own reality. Every day we choose, consciously or not, the things we attract and manifest in our life: happiness or suffering, success or failure, prosperity or poverty.

Some scholars and scientists believe that quantum laws can affect not only the subatomic world, but also our daily lives. We can study these laws to dramatically improve our lives and create a better reality and achieve our life goals.

LIFE GOALS AND QUANTUM STRATEGIES

Consider the electron: In quantum physics there is a process called the "collapse of the probability wave function." Collapse may be understood as a change in conditional probabilities due to the inference of the observer with respect to the observed phenomena. It means that the act of observing a certain particle carried out by the observer modifies the structure of the wave. So the electron appears, out of a pure energy state, as a particle of matter in a certain position between many possible positions because of the observer's action.

Ultimately, what is thought, but energy in wave form?

We can consider a wave function related to every potential event and to the probability a certain event will occur, not just in the life of an electron, but in our life as well. In this way, we can start a fascinating journey toward the "theory of everything."

A quantum approach to daily life suggests that using your brain and the mind, its software, you can shape reality by attracting to you those things you need for a better life. If you can intensely focus on what you really want and keep your mind concentrated on what you have decided to manifest in your life, those things will come to you. It's not easy and requires the ability to maintain your mind at a constant high energy level. You have to inhabit a lucid, sharp state of presence and awareness to be the director of the movie of your life.

There is an open debate between quantum scientists to decide if there could be a possible impact of quantum mechanics on our daily lives. If you speak with most physicists they would tell you that it's absolutely crazy to think that you can intentionally cause the collapse of any kind of wave associated with your thoughts. They state that only the subatomic world is ruled by quantum laws. However, it seems that many prominent scholars and scientists attempting to develop a quantum philosophy disagree and are supporters of the possibility that quantum laws can also affect our daily lives.

If you read books written by contemporary scientists, such as Fritjof Capra's *The Tao of Physics* or Amit Goswami's *The Self-Aware Universe: How Consciousness Creates the Material World*, you can take a glimpse between the folds of reality and gain a better understanding of the principles of how Quantum Trading and quantum philosophy work.

If you are a physicist and you observe a probability wave of an electron and measure it, all of the potential events collapse into the unique, final event. It is similar to what Schrödinger points out with his cat in the box paradox, as we have already discussed in Chapter 1. The cat is both alive and dead at the same time, until you open the box and observe it.

Anything you can imagine potentially could be associated with quantum probabilities. You are the observer who creates the wave function representing your consciousness and thoughts. Consequently, awareness collapses into the particles composing what we usually call reality. As John Hagelin, a contemporary scholar, points out:

According to what quantum physics teach and quantum cosmology confirms Universe essentially emerges from thought and all of this matter around us is just precipitated thought. Ultimately we are the source of the Universe, and when we understand the power directly by experience, we can start to exercise our authority and begin to achieve more and more.

It seems that we are the creators not only of our personal destiny, but we could also be the creators of a universal destiny.

It is no wonder that the use of advanced mind techniques, based on quantum philosophy quite new in the West but very old in most parts of Asia, can cause many positive changes in your life. You can reduce stress, remove mental and emotional blocks that interfere with success, develop a more positive attitude, enhance creativity, improve relationships, and boost your motivation.

Science fiction or quantum reality? It's difficult to decide using only linear patterns of thought, which are very useful in solving daily problems, but inadequate in explaining the nonlocal relationship that exists between energy, matter, consciousness, and reality. What a big problem, but also extremely fascinating!

Though it has been mentioned in Chapter 1, it's crucial to remember that Richard Feynman, one of the greatest twentieth-century quantum physicists, stated that no one fundamentally understands all the implications of quantum physics, even though there are several nice models available, as well as elegant equations that make quantum mechanics consistently work.

Everything depends on how we use our thoughts. How we focus our brain activity directly affects and tunes the neurotransmitters secreted by the brain. Neurotransmitters are endogenous chemicals that relay, amplify, and modulate signals between a neuron and another cell. They influence our moods by increasing happiness or sadness, motivating us to achieve our goals, or amplifying the frustration that causes us to give up.

Neurotransmitters in turn stimulate our endocrine glands to produce a certain level of various hormones, the most powerful medicines or poisons that our body can produce. We can control their production by controlling our mind and our thoughts. Also the food we eat can affect our behavior and emotions. For example, often the food and drink we consume contains too much sugar, causing not only health problems such as obesity and diabetes, but also weakening our character, inclining us to be too self-indulgent and too easily give up. I eat everything, but when I have lunch or dinner I prefer to tap into some special food combination to enhance my energy and health and stimulate a good hormonal balance.

Just as food nourishes the body, thoughts and emotions nourish the mind.

Depending on the focus of our thoughts and emotions we can attract beautiful or ugly things to ourselves.

People usually complain that someone else is the cause of their sorrow, disappointment, anger, and bad luck. Instead of taking responsibility for their own life, they blame others. They are not interested in real life change. They are only interested in sticking to their frustrating life, and they repeat the same behavioral pattern every day.

They continuously find excuses for their fear and confusion. This is the typical mind structure of an eternal loser who wins all of the worst things he can experience. A born winner, whatever happens, takes responsibility for his life and does not waste time and energy blaming others. He doesn't complain or search for excuses. Rather he takes action to transform a problem into an opportunity to improve his life. He focuses his mind on pleasant emotions and he uses them to drive him toward success. He enjoys his life without focusing on fear and guilt, the feelings most responsible for unhappiness and failure. If you can get rid of fear and guilt you will enjoy a meaningful and beautiful life. It is, however, not that easy because contemporary Western culture is based on feeling regretful and sorry for yourself. Please notice that a sense of guilt is one of the most powerful instruments in mass mind control. If you want to be free you have to struggle, but believe me, it is worth it!

Condescending and self-indulgent people are actually involved, through their attitude, in manipulating others to gain their attention and favor, having chosen a weaker way to live. Giving up guilt, condescension, self-indulgency, and fear means renouncing manipulation of others and ourselves, unleashing power from within and opening the door of real freedom. Look within yourself and search for your truth. Jesus taught his disciples, "Then you will know the truth, and the truth will set you free."

Good or bad luck does not exist, or rather, it depends on the way we use our mind at a very deep level even if we are not aware of it.

ARE PORSCHES AND SUPERMODELS EVERYTHING IN LIFE?

If you are able to forecast when a stock or a currency are about to reverse their trend it means that, if you trade the market accordingly, you can significantly improve your quality of life, and have more freedom and time to develop yourself. You can finally choose the life you want and be your own boss.

If Quantum Trading works so well, then why am I sharing this information with other people?

Besides making money to live according to my high standards and to afford nice things like Porsche and Jaguar cars, I like reading, writing, traveling, researching, teaching, and communicating to people my discoveries and experiences. Interacting with people is always exciting, and while you teach you often learn something new about the things you already know. So that's why I like teaching and writing. I always need new stimuli and I don't like chemical stimuli coming from drugs. I don't want to be rich and bored. I prefer being rich and having a lot of fun!

I confess: Besides nice cars, I also like supermodels and going to parties where they also hang out. You remember the supermodel Adriana Lima in one of last year's Super Bowl commercials? Wow! Yet for a gentleman, cars and supermodels should not be everything.

When I was younger I lived as *if today were my last day.* As time goes by I've become wiser and I live as if *tomorrow* will be my last.

Nevertheless, there are many other significant things in life besides beautiful girls and cars, such as exploring the depth of the mind, achieving an emotional balance, and reaching a higher level of consciousness. That's why when I was younger I spent quite a lot of time traveling throughout Asia and I have visited India, Nepal, China, and Tibet. Over there you can learn a lot about yourself, your mind, and the universe.

So far I have written two other books about trading that were published in Italy, my country of birth. I am fascinated by the quantum physics approach to reality, and trading represents only one application of that, though it's a very useful one for making money and enjoying an exciting life.

HOW DO YOU PLAY THE GAME OF LIFE?

So, are you still there? Before starting with some other technical aspects of Quantum Trading, let's play an interesting game: the game of life and its strategies.

Stop for a moment, sit back, relax, and answer the following questions:

- Do you like your current lifestyle?
- Do you have enough time for yourself, your family, and your loved ones?
- Are you satisfied with your quality of life?
- Are you satisfied with your significant other?
- Are you satisfied with your income level?
- Did your investments in the past three years earn good returns?

If you answered "yes" to all of the above questions, I would like to congratulate you! You are a skilled player of the game of life and it's not really necessary for you to continue reading.

If, however, you answered "no" to two or more of the questions, then this book could be very useful. For some, the game of life is mysterious and you believe that some people are just luckier than others. Maybe you're already good at earning money with your professional activity, but the return on your investments could improve. Most likely you've lost money in the stock markets during 2008 and 2009.

Perhaps you want to change your life and you're searching for other options. You want to be financially independent, but you haven't found the way to achieve what you want yet, and you are not yet familiar with Quantum Trading strategies. These strategies have enabled many people, the students who have taken our classes in Switzerland, to consistently earn money in financial markets even when the stock market falls, by trading futures, options, and forex (FX), and keeping losses under control with stop-loss orders.

In the next chapters you will find out that with trading options you can earn anywhere from 30 to 100 percent, and even more, of the option premium within a few days, in the case of a big market movement or change of implied volatility. Yes, you've read that correctly; there are no typos. To do so, however, you must keep risk under control, cutting losses immediately if the market goes in the opposite direction beyond a certain range; otherwise your dreams could turn into a nightmare.

If you choose to trade futures on commodities, stock indexes, and currencies or just stocks, and you always place a stop-loss order, you can achieve your life goals if you have a good trading strategy. If you don't remember to use stop-loss, sooner or later you will lose everything. Stop-loss is your insurance policy to stay alive and in good health in the very dangerous environment of the jungle of trading.

BUILDING POWERFUL STRATEGIES IS THE KEY TO SUCCESS

I strongly believe that trading is a metaphor for life.

If you follow the right strategies, you will be very successful. According to Neuro-Linguistic Programming (NLP), having the right strategies is the key to success in life. According to this discipline, losers follow the wrong strategies and refuse to change. In the end, they say they are unlucky....

NLP studies how people create strategies to organize their thoughts, feelings, language, and behavior to achieve results, even if they themselves are unaware of the process. NLP teaches people how to make big changes in their lives. It proposes a methodology to successfully model the outstanding performance of geniuses and leaders in different fields.

This means that you can achieve extraordinary success if you are able to create a consistent and "congruent" strategy to achieve your goals. NLP cannot create miracles, but it can help you to understand how to build an effective strategy to achieve the best results in the shortest amount of time.

NLP advises us to study how the most successful people, in the field you're interested in, have achieved their position. You should study their strategies and divide the entire process into a series of concrete, accomplishable steps. Then you need to clearly consider how to model your behavior, following a similar strategy. So you begin to take action step by step. As you move from one step to the next you have to check if each step is giving you the expected result. The final goal is eventually achieved when each intermediary step is successfully achieved. "Congruence" between one step and the next of your strategy is the key. Feedback process and control of the result of any step of your strategy are mandatory. In case the result you obtain from a step is not satisfactory, you need to come back to the previous step and correct the action related to it until you get the expected result. This process is not a cakewalk and most people give up and complain that they are unlucky, or they keep on making the same mistake without asking themselves what is wrong. Luck is about life change and committing yourself to following all of the steps and never giving up. If you fall down, take a rest, ask yourself what went wrong, and then focus on correcting your strategy to get the result you desire. No pain, no gain.

A key concept of NLP is that we form our own unique mental maps of the world as a result of the way we filter and perceive information absorbed through our five senses. Even if the sensory input and the information is the same, people process the same data using different sensorial strategies, obtaining different results. If each time you have tried to achieve your goals something has gone wrong, it means you are using ineffective strategies or you are repeating the same mistake. What we are used to calling reality, then, is not an objective entity. When we look at the world we are actually just selecting one way out of many to represent our experience. Reality, then, is only a mere representation. It is one possibility out of many, like the position of an electron that is found in a precise position only when observed, but that could actually be anywhere within the probability curve function, as Eisenberg and Schrödinger pointed out.

NLP techniques are very powerful for designing effective strategies to reach the best results in fields such as mental health, sports, politics, business, motivation, and trading.

NLP teaches us how to analyze a system of relationships and transactions, reduce it to its basic components described by sequences of representative sensory units, and then work with it. These techniques have become very famous within the past 30 years. They are used by the most successful people around the world to improve performance throughout various sectors.

I am positive that NLP can be useful to traders using various discretional approaches. They would find it very beneficial for improving the application of their money management strategies, or positively conditioning their brain to imitate the top traders' risk management strategies. If, when you trade, you often lose what you've gained, your risk management strategy is probably not very good and you need to change it. When I teach my students Quantum Trading strategies, at the end of the course we spend several extra hours on trading psychology, NLP techniques, and much more. Trading psychology is what makes the difference between an excellent trader and a mediocre one.

Quantum Trading techniques can offer you the basic components with which to build up various effective strategies to be successful in trading and obtain wealth and freedom.

If wealth alone cannot bring you happiness, at least it can help you avoid many of life's annoyances. But what is real wealth? Wealth, for me, is just a tool to increase quality of life. Wealth itself is not the final goal, but merely an instrument for increasing your freedom.

It is very important to have dreams and life goals. Wealth can help you achieve them; yet, you also need a life-philosophy, some model that inspires and empowers you to obtain what you want to achieve. If you don't have dreams or goals that drive you, you will find it difficult to reach a significant level of wealth. Even if you do, you could lose your money sooner or later, or even worse, lose yourself. Intention, awareness, and knowledge are more important than objects that you merely own. If you continuously improve yourself and aim to reach a higher and higher level of knowledge and awareness, you will exponentially increase your personal power and improve your energy level. You will take a life quantum leap. Then wealth will come to you like a moth to a flame.

CREATING MOMENTUM IN LIFE

It is indeed possible to start from scratch, be successful, and achieve a lifestyle where you have everything you need to feel happy and fulfilled.

The key words are "no limits." I especially refer to your mental limits bounding people to a class B life. These boundaries are our beliefs, the most powerful software units of your mind.

Tony Robbins observed during one of his seminars that, "Beliefs have the power to create and the power to destroy. Human beings have the awesome ability to take any experience of their lives and create a meaning that disempowers them or one that can literally save their lives."

So, your life-potential is determined, or limited, by your self-belief. Our values and beliefs shape our actions. Anything is possible if you focus on passion and purpose.

This is what NLP suggests. Many examples of self-made millionaires confirm that.

It's very important to always check which beliefs guide your life and if your belief system is able to empower you to achieve your life goals. If not, then you had better change it.

A good strategy for success begins, like Paul Getty suggests, by "doing business with our own selves." In other words, it's very important to start by investing in ourselves and find an activity that permits us to exponentially increase our profits in relation to the time dedicated to the business until a critical mass is reached. Once you have reached a critical mass you can dedicate more time to doing what really interests you. So choose an activity that carries *momentum,* that is, a powerful shove in the direction of continuous growth and development.

If you are ambitious and feel you have many goals to reach, and yet within five years from now you are earning about the same amount that you earn now, something is probably wrong! Your life does not have momentum. Your growth is too slow and you don't have the potential to reach your top goals. If true, you will have to be content with being a spectator to your own life.

A great example of momentum comes from the biography of the legendary trader W. D. Gann. Coming from very modest origins and lacking any financial resources, he dedicated 10 years of his life to studying stock market charts in search of a cyclical law that could enable him to spot tops and bottoms of a trend. His love and passion for work along with his ability to focus 100 percent on studying hundreds of stocks and commodities charts led him to discover the "vibration law." Using this knowledge, he created techniques to analyze the financial markets and trading strategies to consistently beat the markets during his 45-year career on Wall Street. During this time he allegedly made a fortune of over $50 million.

During his career Gann traveled with his private plane and loved to pilot it himself. He was one of the most respected personalities in the financial world. His life was filled with travels around the world where he studied ancient civilizations' wisdom in geometry and mathematical systems. It seems he was also a good friend and an advisor of the Rockefellers. He was inspired by the mathematical structure of the Great Pyramid in Giza to create very precise forecasting techniques. Gann wrote many books about trading, but he never commented publicly on his more secret techniques; those were only revealed to an inner circle of students and friends. He had a fear of being considered crazy by his clients, so he wrote only general, but accurate, books on trading concepts.

Even though this is not a book about Gann, we'll explore in the next few chapters some of Gann's approaches to the market. He was a visionary who, by 1908, had already thought of trading in terms of atoms and electrons, as he stated in a famous interview with *The Ticker Digest*. His existential path was oriented toward continuous growth. He lived a long and enjoyable life. Shortly before dying he wrote in one of his books that he had lived with fullness, that he felt satisfied, and that he wasn't scared of the last big journey awaiting him because he had obtained everything he desired in life through financial trading. Gann was a real master of utilizing options and futures to create momentum, and his life is a beautiful example of how passion, commitment, and focus can allow you to achieve your goals.

CROSSING THE BRIDGE

Let's begin this section with another couple of questions. Questions, according to NLP, are very important because they provide you with the key to unlocking your unlimited potential.

- How much time do you dedicate on a weekly basis to asking yourself if you are satisfied with what you have and if you're on track to reach your goals?
- How much time each day do you dedicate to finding strategies that can improve the quality of your life by working less and earning more?
- Are you sure that today's certainties will be here tomorrow? It's not a question of pessimism, but rather of realism, similar to the workings of the stock exchange.

By studying history, one can see how the biggest economic and social crises usually occur after financial crashes. The current big recession in the United States, stemming from the subprime mortgage and toxic derivatives crisis, has caused thousands of people to lose their homes and their jobs. Unfortunately their suffering speaks for itself.

Throughout the course of this book we will introduce very powerful trading strategies that will enable us to understand how it is possible to program constant returns in our trading, even if our starting capital is relatively small.

Before introducing our Quantum Trading strategies in the next chapter, we should take a moment and reflect on what it is that both influences and determines the results in our life. I believe this is necessary to attain the maximum benefits from these powerful instruments, which can generate a lot of wealth if used correctly.

Many people in the world believe that it's very difficult to attain what makes them happy because it usually does not happen for the average person. "Reality is really different," some say, "and it's not easy to get what you really want. If it were easy, everyone would be rich and happy." Their mind set simply doesn't allow them to think differently.

But are we sure that this is how it really is? There is a bridge in your life, beyond which exists a life of liberty and financial independence. Some might call it success; however, few people actually dare to cross this bridge. Most people are afraid of the risks they could take in crossing the bridge. Some even think it's prohibited to cross the bridge. Only once in a while does someone attempt to cross the bridge, even though everyone else thinks she's crazy and laughs at her. But when that person succeeds and reaches the other side of the bridge, the others say, "Look at her and what she did. How lucky she is to have become that successful!" These people are the ones who will never dare to cross the bridge and will instead stay where they've always been, complaining about everything.

There is also another type of person who says that he wants to cross the bridge, but not now: he doesn't feel like it today, but perhaps he will tomorrow. This person apparently contemplates the crossing, but is always indefinitely postponing it.

The lives of W. D. Gann and other people who have successfully crossed the bridge prove to us that it is possible to achieve what we want if we develop the right focus and direction.

Where is this bridge? It is a place in our minds that I hope you can cross as soon as possible, if you haven't already done so. And what is it made of? Perhaps Shakespeare would say that it's made up of the same material as our dreams.

I prefer to say that the bridge is composed of our belief system, which molds the way in which we see the world. What empowers us to cross the bridge are our values and taking a congruent, effective action.

We don't actually see the world for what it is in reality. We see it from our point of view and through our structured belief system. Our beliefs are the filters that we use to interact with reality. People always see reality through filters and "the map is not the territory," as the Polish-American scientist and philosopher Alfred Korzybski remarked.

We have to carefully examine our deepest beliefs and values. You will discover that often we are not fully aware of them. Beliefs and values are the compass that we use to direct our lives. To cross the bridge and achieve success we have to cultivate the right beliefs and values and change the ones that are not beneficial. If your beliefs tell you that it's possible to make a lot of money and enjoy life, then you will pursue your goals, in spite of obstacles, because you're sure of yourself: Your internal image of yourself is very powerful and motivating.

If, instead, your beliefs tell you that only those who are born privileged, or who steal, or who use unethical means to achieve the ends of wealth can become very successful, then you are bogged down in a murky swamp. It's time to start reorganizing your values and beliefs before making any other changes, or pursuing financial trading.

Now I have some other questions for you.

Take out some paper and a pen to write down your answers clearly and concisely. It is important to write your answers out because thoughts are often volatile and we're not used to systematically questioning our existential goals. Writing helps us organize our thoughts and pinpoint the answers. Do you remember? Asking the right questions helps you to unleash your power from within.

- What beliefs guide you in life? What do you really want to achieve in life? Look inside yourself and think of the most beautiful things on earth you desire. What are your beliefs about your ability to achieve what is really important to you?
- How would you like to live your life? What kind of person would you really like to be? What makes you happy? In other words, what fulfills you?

Now that you have focused your life goals, you need to review the strategies you intend to use to accomplish them. Perhaps you have already thought about what you want to achieve, but now you have to ask yourself if your beliefs and actions are in line with your goals, so please answer the questions below:

- Does your belief system allow you to live a first-class lifestyle?
- Are the strategies that you are currently using to manage your financial and professional life able to bring you everything you desire?
- How much have you improved your economic level, and what have you achieved compared to what you planned to achieve in the past 12 months?
- If you are already a trader, does your money management strategy and trading system help you accomplish your previously set goals?

So many questions! Isn't this supposed to be a book about Quantum Trading?

Don't worry; I consider this a preparation for developing a strong motivation and achieving extraordinary results with Quantum Trading! Using powerful instruments, like futures or forex spots, without having these top goals in mind, would be completely useless.

If you are totally motivated you can achieve really impressive results. Dedicate at least half an hour every day to orienting your life in the direction that you want it to go. Think of all of the most beautiful and motivating things that turn you on, feel the emotion associated with these images, and you will create a magnetic field of attraction. The secret to making everything work is creating a clear image of what you want, using vivid colors. Then, carefully examine the images as if you were projecting them onto a movie screen. Try to imagine the sensations and emotion of accomplishment and associate that feeling with the images as if they had already happened. And then take action. If you don't act, then you're only daydreaming.

If you have a good understanding of your mind, the thoughts and the mental images that you create will become reality, *nous*, and will develop an Entelechy—or rather, the process that perfects itself as it occurs—as Aristotle used to say.

When I spent several months in Tibet to study the structure of the mind, the first thing I was taught was how to observe the mind and the different structures of thought. Have you ever tried to be mentally quiet for one consecutive minute? It's almost impossible the first time you try. You need a lot of training and practice to control your mind.

What I want to suggest is not to switch off your brain, but instead to examine your thoughts as if you were an external, attentive observer. At first it's nearly impossible to do so, because you develop an acute observation of mental space, where you will find an endless flow, a chain composed of thousands of thoughts, streaming through your mind. At the beginning you will notice only the most prominent thoughts. As your mental presence strengthens you will find that there are subtle thoughts of which you are normally not conscientious. To identify their content you have to create a very specific focus. These thoughts flow unrelentingly and establish the deeper structure of our mind, which in turn generates our reality. The mind, according to ancient Tibetan wisdom, is *Kunjen Gyalpo*, "the king that creates everything."

Western neuroscience and contemporary schools of psychology today confirm what has been known in Tibet for thousands of years. Transactional analysis, Gestalt principles, and NLP point out how we unconsciously accompany our decisions with a continual, subtle flow of comments and internal images, which drive our motivation for every action, important or banal. They are the underlying structure of our mental, emotional, and psychological reality. Whether or not we are aware of them, they are what determine who we are and what we will achieve.

If you pay close attention to your thoughts for awhile, you will discover the true and proper structures of your *metacomment*. Without delving into deep psycholinguistic analysis, we can say that some people possess a set of positive metacomments. When they start something new and encounter obstacles, their inner voice tells them, "Come on, keep going, you can do it. Stick to your decision and get what you want."

Others possess a set of negative metacomments. Any obstacle in the way of their goals, because of these metacomments, transforms into an impossible barrier: "Just drop it. If it were actually possible to achieve total financial independence, then everyone would succeed," or "Yeah, I want to start this new process in my life, but not now, maybe next month when I'll have more time." They never get around to starting, or they continuously change their mind and abandon any effort because they feel discouraged. A determined willingness, applied to strategies you decide to put into action, can bring you outstanding results.

In any case it's important to frequently evaluate and monitor your strategies. In trading, an excess of trust or distrust can be fatal, especially if we don't understand how to create an efficient strategy by following the rules we were given at the beginning and applying them.

To be successful it's not only necessary to have the right thought process, but you also need the appropriate strategies to realize your goals. You must monitor this process continuously to evaluate if you are applying the strategies appropriately.

Congruence in taking action means success. Incongruence means failure. Quantum Trading is a congruent system that allows you to be successful in trading. Anything is possible if you focus on passion and purpose.

CHAPTER 3

Quantum Trading 101

T echnical analysis is a really fascinating discipline. It can help you understand the trend and the main figures, like flags, triangles, Gann's and Fibonacci's retracements, and double and triple tops and bottoms. It is also very useful for spotting likely price levels of resistance or support where you can open a new trade. This chapter covers some of the basics of Quantum Trading, outlines some good rules to follow, and introduces a Quantum Trading model for further discussion.

A GLIMPSE AT SOME TECHNICAL ANALYSIS PROBLEMS

If you have studied moving averages, oscillators, and indicators, some of the most peculiar technical analysis tools, you immediately realize that they are full of unconnected and conflicting techniques. For professional traders and investors it is often difficult to select and harmonize different technical tools because they frequently show contradictory results.

For example, display on your favorite stock XY chart a simple technical tool, such as a 28-days simple moving average (SMA 28). Assuming you want to buy your favorite stock long, you wait for the price to cross the SMA 28 to buy, and you sell it when the price crosses the SMA 28 going down. You can also use the crossing of two moving averages as well as add some other filters. It is easy to make money when the trend is steady, but if the price moves into a lateral trend or the trend is choppy you can lose a lot of money since you can get many false entry signals.

And why did you choose the SMA 28 for your favorite stock XY, and not some other SMA, such as the SMA 14 or SMA 21? Probably because you have done the back-test before choosing the 28-day period as the best fit for the past three to five years. You have tested your stock XY's historical data and you have done some data-fitting to optimize your result. The only problem with this process is that 28 days for your SMA is the best period you found by fitting the XY stock with past data. Unhappily, when you start to trade with real money you will quickly realize that the past is a different story from the present; you will experience some unexpected drawdown and lose a lot of money. Why? Because the trend structure is nonlinear and much more complex than a data-fitting process can take into consideration. As Jim O'Shaughnessy writes in *What Works on Wall Street*, "Torture the data enough and it will confess to anything." I like to back-test, but I am always very careful with the results.

Let's figure that you want to limit drawdown, and therefore you filter the signal you receive from your SMA 28. You try putting together your SMA 28 with some of the other many indicators and oscillators available in the technical analysis arsenal. But doing it this way makes your life harder than it should be. In fact, you have to answer several additional questions, such as "Why use a moving average convergence/divergence (MACD) filter instead of a relative strength index (RSI) one, or a stochastic indicator?" or "Which are the best two indicators among all of the different tech tools?"

Assuming you are skillful or lucky enough to choose the best indicators, when you receive heterogeneous or contradictory signals, as each tool is based on different principles, you will become even more disappointed.

However, you're probably another kind of trader. You are not a trend follower and you love contrarian trading. So, you prefer trading with overbought and oversold indicators. If they show you that they've reached the red flag area for a reversal and the trend keeps steady and strong, you are in big trouble and risk losing a lot of money. This is because your overbought indicator shows you that you are about to have a top and you should sell your XY stock short, but instead it doesn't want to plunge and price keeps soaring.

The same thing happens in the case of being oversold. You lose money as you continue to buy in the belief that the price is ready to rebound. That's what happened to many stock indexes in 2008 and in the first quarter of 2009.

The Quantum Trading approach instead taps techniques and tools all generated from the same viewpoint. If you study them deeply you will find harmony, elegance, and logical coordination.

Consider now one of the easier technical analysis concepts about supports and resistances and their effects on price level: You buy on a support and you sell when price reaches a resistance. You spot the A and B points on your chart and you wait for the price to touch point C to buy and open a long position.

Take a look at your favorite stock or commodity. You can easily spot double bottoms on the same price level, or if you want to use some dynamic supports you can draw a trend line joining two higher bottoms at point A and point B. In both cases you decide you need a third point as a buy signal. Therefore, you wait for point C, which is another third-higher price point level on the same trend line. Or you can wait to buy your stock, until the price comes back to point C on the same price level of a double static bottom, hoping it will be a triple bottom.

Now take a look at Figure 3.1. If you want to use point C as a buy signal, you are supposed to wait first for points A and B to show up in the chart. Then you join them with an extended trend line and wait until the price comes back onto the trend line. The point where the price touches the trend line is C point, where price, in this case, rises up.

Point C obviously shows up only after that the first and second bottoms have already formed. So you need two points first (A and B), and then you have to wait for the third one (C).

FIGURE 3.1 CME Group EUR-USD Futures Chart: The C-Point Buy Signal on a Trend Line

MAKING MONEY WITH QUANTUM PRICE LINES

Quantum Price Lines (QPLs) can show you the level where you will most likely find point A—that is, the first bottom or top—while it is forming on the chart and without waiting for the appearance of points B or C.

In other words, QPLs can predict the most likely levels for a major reversal. For example, Figure 3.2 shows the QPL predicting the historical top of the S&P 500 index future. Notice the dashed lines in the figure: those are QPLs, one of the most powerful tools of our Quantum Trading arsenal, representing crucial points of support and resistance. If you are an advanced quantum trader you will discover that some of these lines are specialized to offer support resistance levels, while others are more likely to show very strong support.

In any case, at this early stage we'll view these indicators as simply as possible, without trying to distinguish between ones more likely to be a support or resistance. We'll consider them as if any line could offer a very strong support or resistance level.

On July 16, 2007, the S&P 500 future price touched the quantum line at point A and immediately after we saw a very big drop of the stock market. The price bounded back up on point Y on August 16. I will never forget that day.

FIGURE 3.2 S&P 500 Future All-Time Historical Top with Quantum Price Lines

During the night S&P 500 index prices were still plummeting after more than three weeks of consecutive losses. The market opened with a big gap just on the lower Quantum line price on point Y, able to offer a great support. I had been waiting for that moment for several days. I had just bought long the future before the opening pit session that same morning, when, after a few minutes, Jim, a close friend of mine, arrived at my place to have breakfast before leaving for Los Angeles. He asked me what I thought about the stock market, and when I told him I had just bought the S&P 500 index future he observed that I had to be crazy to do such a thing, because the market was collapsing and nobody knew when it would stop. I told him that I believed it would imminently rebound. He asked me how I became insane enough to believe such a thing; Jim is just a bit radical. When markets go up he thinks it will last forever, and he becomes a pessimist when they go down.

I told him that I was buying the stock index because the price was just on the QPL at 1393 (point Y) and that the same quantum line, on March 6, 2007, had offered a very strong resistance at point Z.

I had already placed a limit order on that level.

When the prices continued to plunge, Jim seemed very worried about the markets. He kept watching the screen.

"Hey, man! You have bought long the S&P 500 future, and stock prices keep on going down!" he exclaimed, while I prepared a very promising dish made with Greek-style yogurt. "Let me see what the folks at CNBC will say about that!" he added, picking up the TV remote. As I placed honey, cinnamon, a little muesli, and fresh cut fruit on the top of the yogurt, Jim asked me with surprise, "Fabio! How can you be so focused on preparing breakfast when S&P futures are so volatile! Didn't you recently open a long position?"

I replied that I had placed a stop-loss order at 1373.00, just 20 points below the bottom, and if it was kept I would only have lost what I had established before opening the trade. Statistically, using my Quantum Trading tools, the stop-loss is caught three out of 10 times, but I make money the rest of the time.

I usually place a seven-to-10-point stop-loss on S&P 500 futures for swing trading in normal market conditions and a stop-loss of only three points for short-term trading. But that morning I had used a 20-point stop-loss because volatility had dramatically increased during the previous days, and put and call showed unbelievable prices. When such a thing happens you need to use a larger stop-loss because volatility is one of the most important factors in calculating your stop-loss, after establishing your maximum risk and loss tolerance for each trade. This makes your trading more effective, especially if you use a mathematical model to adjust stop-losses consequent to volatility changes.

If volatility increases dramatically, you cannot use the same stop-loss size you used before. By not taking a higher risk you can use a bigger stop-loss and reduce the number of contracts you are trading.

I was expecting a decent bounce in the next few days before closing the position. I would keep my long position for about 100 points, in any case for not more than a week, for the trading time window that I was trading.

When the price of mini–S&P futures reached 1377, Jim told me that my stop-loss was likely to be caught. I suggested that he taste the yogurt and tell me what he thought, instead of worrying like that. I had just discovered a product that is the original Greek yogurt, which has a rich taste, only after much research, and—believe me—it really made a difference.

After you have placed a stop-loss you cannot sweat out the trading in front of the monitor, getting nervous each time the price goes in the opposite direction of your trade.

Trading is not gambling and, instead of allowing the market to stress me out or nail me to my chair, I prefer doing other, more exciting things. Once I have placed a stop-loss and a target exit price, I can relax.

Then something unexpected happened. The S&P 500 future price fell to 1375 and my stop-loss was about to be caught when suddenly great news popped up. It was reported that the Federal Open Market Committee (FOMC) meeting had resulted in the Fed cutting interest rates. It was really astonishing. While the market was severely plunging, Mr. Bernanke had done his dirty work, and the stock market reacted immediately. The S&P 500 future price soared 50 points within a few minutes.

Wow! At that time Jim didn't know that soon interest rates would approach zero. Mr. Bernanke would not have any more room for rate cuts, and even though I suspected such a thing after taking a look at subprime numbers, I could not be sure at that time. I needed another test of the resistance given by the QPL, which the market had topped some weeks before at point A. If price proved unable to break this QPL of higher resistance then we would see another top, from which a long and severe correction would begin, according to my Quantum Trading time algorithms.

S&P 500 futures were rallying so strongly that Jim couldn't believe his eyes. He stopped eating and he didn't say a word for the next ten minutes. Not only had the price come back to my entry price, but I had gained about 30 points in the time I had enjoyed my delicious breakfast. When you place an entry order using QPLs, you always have to place a stop-loss. This allows you to stop worrying about the price going up and down.

In the beginning it can be difficult to distance yourself emotionally, but after many years you get used to it, and you are more detached from market intraday volatility. You become interested only in whether or not your stop-loss will be caught. If not, your position is still open and your chance to make money is still there. This happens to skilled quantum traders about 70 percent of the time!

Now I have some questions for you. Are you ready?

Why did the S&P 500 price rebound so strongly after making a magnificent spike bottom so close to the QPLs? If you want to be very precise you need to consider that the slippage was only about 1.45 percent and the potential range of the upside, according to Quantum Trading techniques, was 167 points (that is the distance between the two consecutive quantum price lines where you find points A-B and Z-Y).

Was it only a coincidence that Bernanke had announced an unexpected and dramatic rate cut just after the S&P 500 price touched the quantum line support?

If not, then who had influenced what? Or, what had influenced whom?

Were the two events governed by a cause-effect relationship or should we consider them as coemergent and synchronistic, as Carl Gustav Jung might have taught?

If you intend to find an answer using a linear approach, forget it! Linear systems of thought are unable to deal with this type of nonlinear issue, nor can they solve any problems related to stock-market forecasts that are also ruled by nonlinear functions. If you choose a linear approach, then stop reading, close the book, and follow only stock trading fundamentals for investment decisions, and remember to cross your fingers before buying.

If you are a little more open-minded and you want to learn more about quantum thought, you might be disoriented at first, but if you stay with me this will all start to make sense.

Take a more careful look at Figure 3.2 and you will see that the same set of quantum lines that we used to forecast a major top on point A and the bottom at point Y offer another powerful resistance price level at point B, the S&P 500 all-time historical top on October 11, 2007. Because the price was unable to break the powerful, uncanny resistance in level B, the trend lost its momentum and the price plunged.

But where will the down leg end, and the price find a support? According to my quantum models, the stock or commodities price tends to behave like an electron. If it starts moving away from point B, it can reach point C or point A, as you can see in Figure 3.3, depending on whether it loses or gains energy. Don't worry if Figure 3.3 shows a sketch of an atom similar to the initial Bohr model of the atom, which resembles a solar system. Even though contemporary physicists think the model is outdated, it was really important in the history of physics to display the essential qualities of the atom as intuitively as possible.

FIGURE 3.3 A Sketch of an Atom with Its Orbiting Electrons

Bohr described his model of the atom using the analogy of the solar system. In place of the sun, the system is centered on the positively charged nucleus, and instead of the planets, the nucleus is orbited by the negatively charged electrons. At the beginning, Bohr thought that the electrons moved on an elliptical path around the nucleus, which is composed of protons and neutrons. Coulomb attraction of the nucleus for the electron provides the centripetal acceleration. In this way we can relate the radius of the orbit to the electron velocity. I don't want to bore you with the equation, and so I'll avoid putting it below.

The only problem with this "solar system" atom model is that it is not stable because the accelerating electrons emit energy and so, at a certain point, they could lose energy and fall down into the nucleus in only a few orbits. So the atom would be unstable and what we understand as matter couldn't exist. Bohr solved the problem tapping into Planck's idea of "quanta" of energy, assuming that the orbits that the electron describes are discrete. So the electron's charge remains stationary and doesn't change its energy value circulating on the same orbit, because it doesn't radiate energy (light). It will happen that it will emit light as a quantum only when an electron leaps from a higher energy orbit to a lower one.

Finally, if an electron is found in its lowest energy orbit, there are no lower levels to which it can jump. In this way, the atom is stable and there is no danger that the electron could crash into the nucleus. Another important issue of the Bohr model is that an electron can never occupy any position between orbits. When it takes a leap it seems it moves directly from one orbit to the next. This may seem weird, but it's exactly what the electron does.

This is why in Figure 3.3 we display three possible positions for our hypothetical electrons in orbits A, B, and C (the electrons in Figure 3.3 do not correspond to the actual structure of the atom. It is only for the sake of explanation). Assuming that the price is like an electron, it will move from point B to A or from point B to C. The quantum orbital corresponds to our QPLs.

You can see QPLs both as a manifestation of the point of maximum curvature of P-Space, using a model stemming from the theory of relativity, or as a quantum orbital. In the last case, when we say that the price-electron cannot occupy a position between orbits it means that a stock's price, if it touches a support QPL, like point B in Figure 3.3, can move toward point A by rebounding, or fall down to point C.

The analogy is not literal because, while the electron quantum jumps, it teleports the particle from point B to point A and the price, instead, will temporarily occupy intermediate positions between B and A, but it is attracted toward A or C, considered in our Quantum Trading model as the natural targets for our price-particle-electron.

The electron's movement from B to A or from B to C corresponds to what you see in Figure 3.2. The S&P 500 future price couldn't break the resistance at point B and so moved toward a lower energy level. You can calculate it in advance using the formula I will give you in Chapter 5. Please note that this same QPL, where point S lies, offered a very strong support on point Y some weeks before. This is the point I used to open a long position exactly at the moment that the bottom spike formed on the chart.

Keep looking at Figure 3.2 and you will realize that the price bounces after a failed attempt to reach the higher QPL. Then the price comes back and breaks the support at 1413.50 on the QPL where you find points Y and S.

What is the meaning of this QPL support break? Simply, that the price at 1413.50 is the last opportunity for the trend to bounce back and confirm its bull trend. In case it doesn't, the market is ready to show the beginning of a bear market where the next target will be the next-lower QPL, where you find points D and E around 1262. On this level the S&P 500 shows a double bottom, but we don't need to wait for the second bottom as all other technical analysts need to do. If you know Quantum Trading techniques you can buy long the S&P 500 future on the first bottom (D) and you can exit at the prudential target of 1337, which is 50 percent of the range identified by two QPLs where you find point D and point F.

The price fails to rise up to the higher energy level and cannot again touch the resistance of the higher QPL (F). It stops before reaching level F, a sign of trend weakness. So, it bounces back and finds a support on the lower energy level in E, on the same level of the previous bottom at point D. By touching point E, the price gains momentum and can leap up to point F, just as an electron receiving additional energy does when it is bombarded with Gamma rays. As a consequence it makes a quantum leap and reaches a higher energy level.

After the price fails to break the F-QPL and starts to plunge, it breaks all of the QPL supports one after another until reaching the frightening level at point G. Finally, the S&P 500 index reverses its movement showing a spectacular V bottom—thank goodness!

Point G lies on a very special level where the probability for a medium-term trend change is much higher than the next highest QPL. Why? Because the QPL where you find point G is the "mother" of all the other QPLs. I call it the "Mother QPL" because this is the first point you have to calculate if you want to draw QPLs on your charts. When you find G-QPL you can calculate all of the other ones, which are nothing other than harmonics of the base tone. This line is thicker than the other ones, and you can immediately recognize it on the chart.

If you are familiar with technical analysis you probably wonder when we have drawn all of the QPLs in Figure 3.2. Have we drawn them before or after the price touched each line?

We have definitely drawn them beforehand, and it's actually possible to draw them years in advance of the price even reaching those levels.

How it this possible? It is because QPLs do not need any previous point to join with, as do trend lines or other technical tools.

QPLs are calculated by equations that are shown in Chapter 5. The equations describe the QPL's behavior in P-Space and cover all of the potential prices where stocks, commodities, bonds, and currencies strongly react. P-space is a virtual space where price and time interact to generate all the movement you see every day in the stock exchange.

If you are an absolute beginner in quantum physics, we need to review the concept of the electron quantum jump to better understand our Quantum Trading model. See the box for a comparison of the Quantum Trading price model and electron behavior.

If you have a PhD in quantum physics, then you probably would prefer more equations, fewer words, and also complain about some of the generalizations used; however, for our trading purposes it is better to present the analogy between electron and price behavior in a very simple way. You don't need this information to apply for the next Nobel Prize, but only to make money trading the markets.

Comparing Price and Electron Behavior

You remember that we said a stock price is like an electron that is able to jump from one orbital level to another. Its movement is ruled by nonlinear functions, and quantum mechanics explains how it can move from a certain energy level, which corresponds to a certain QPL in our trading models, to another one that is more or less distant from the center of an atom.

An electron's energy level is the amount of energy required by an electron to stay in orbit. The electron's position in relation to the nucleus gives it potential energy. An energy balance keeps the electron in orbit and as it gains or loses energy, it "jumps" to an orbit closer to or further from the center of the atom.

This distance for electrons is ruled by Max Planck's constant (h), which shows us the minimum quantum distance between the energy levels of two electrons. So when we speak about quantum leaps and quantum movement, it is because this movement can be measured by a quantum-size package.

The four rules of thumb for price/electron movement in a QPL grid are:

1. Reversal model: When price is falling and touches a support QPL, and immediately after reverses its direction and begins to climb, then it gains momentum. The first natural target price is the next highest QPL.

2. Reversal model: When price fails to break through a resistance QPL, but just touches it and then starts dropping back down, it loses momentum. The first natural target price is the next lower QPL.

3. Acceleration model: When price fails to bounce on a support QPL and breaks it by going downward, then price will continue to fall. The first natural target price is the next lower QPL.

4. Acceleration model: When price reaches a higher QPL and breaks through, it continues upward and gains momentum. The first natural target price is the next highest QPL.

TRADING RULES OF THUMB

If we compare the price of a financial security with an electron particle, the price behaves in just the same way. If it receives an energy shock then it will reach the next higher QPL, and if it receives a negative energy shock, then it will reverse its path, going to the next lowest QPL. If it breaks the resistance offered by the next superior QPL, then it's as if it has taken a quantum jump in the direction of the next highest orbit, which you can easily identify as the next highest QPL. Thus the target price will be always be indicated by the next highest or lowest QPL.

My Quantum Trading model provides you with the equations to calculate QPLs, or price/electron orbits, and you can draw them on charts of stocks, indexes, commodities, and Forex currency pairs.

By drawing the QPLs on our two-dimensional graph, we can see how the price of a security resembles the behavior of an electron.

Now we have to explain something crucial about the QPL's features and properties. They are connected to the concept of quantum leaps. As we have explained, stock, commodity, and currency prices behave like an electron jumping from one energy level to another, that is, from one QPL to the next one.

Carefully consider the next few statements about a QPL's main features and trend lines (TL). For starters, QPLs can help forecast the next top or bottom. Using advanced Quantum Trading techniques and working with P-Space algorithms will allow you to explain in advance when and where followers of technical analysis will see a major or intermediate top or bottom on a particular price level is likely to form, before it forms on the chart.

If you are used to working with many indicators such as stochastic or MACD and you do not find them in the pages of this book, don't get upset. I have too many new things to show you about QPLs and other tools. You can study dear old technical analysis in many excellent textbooks. For this book you will study the Quantum Trading approach and learn how to forecast major and intermediate reversal prices of stocks and commodities in advance.

Quantum Trading Model Facts

1. Each QPL can be a very strong support or resistance level on which price reverses or accelerates its speed beginning a quantum leap. QPLs correspond in some way to the various energy levels of an electron: that is, the position it occupies according to its probability wave structure.

2. QPLs can be drawn on a chart even weeks or months in advance by using the algorithms that we explain in Chapter 5.

3. QPLs don't need previous points to be calculated and to be drawn on a chart.

4. QPLs can predict spike tops or bottoms contrary to other techniques such as trend lines (TLs) where you need at least two previous points to join up and generate a third entry point.

5. TLs often lie on the same price levels indicated by QPLs.

6. TLs can be considered generalizations of QPLs and the latter can generate the former, even though TL users are not aware of this fact.

7. QPLs show other important price entry levels that TLs, Elliot waves, and other indicators cannot show.

8. Some points on QPLs can be very near classic double tops/bottoms, but other points offer unconventional turning point levels where very often the market reverses its trend.

9. Time acts as a trigger for reversal on a QPL, but time quantum techniques are too advanced for this early stage (see Chapter 11).

10. Quantum Trading doesn't deal with deterministic models, but it does assign several degrees of probability to the various QPLs and Time Quantum Techniques in order to work only with the combination showing the highest probability of a reversal.

11. In case an expected reversal fails to occur and the price breaks the QPL, you can decide, under certain conditions, to use a stop-and-reverse order instead of only a stop-loss to close the old position and open a new one in the same price direction.

CHAPTER 4

Sun Spots, Geomagnetic Storms, and the Stock Market

S itting at home one morning I received a call from an old friend from New York. "Ciao, Fabio! This is Charlie," I heard through the speaker. He continued, "I just received this article from a colleague of mine at the Federal Reserve and I immediately thought of you and your research. Believe me, it's really strange! I never thought that the Fed would study such a topic."

I was intrigued. "Thanks, Charlie," I replied. "Send it over and I'll look it over."

My friend's certainty that this article from the Federal Reserve Bank of Atlanta was relevant to my research was correct. The article was titled, "Playing the Field: Geomagnetic Storms and the Stock Market."

The article was fascinating because it supported my theory of the existence of a big entanglement, or a nonlocal giant field, including our solar system that affects the prices of financial markets, or, in other words, shows a correlation with stock prices. The Federal Reserve Bank of Atlanta published this study by Anna Krivelyova and Cesare Robotti a Working Paper 2003-5b in October 2003. Titled "Playing the Field: Geomagnetic Storms and the Stock Market," the paper can now be accessed on the Web (www.frbatlanta.org/pubs/wp/working_paper_2003–5b.cfm? redirected=true).

It seemed that someone at the Fed thought that the activity of celestial bodies like the sun, which causes geomagnetic storms, could in some way affect stock market returns. Or at least someone believed there to be some level of correlation between these two classes of phenomena. Can you imagine that? Serious bankers in pinstripe suits and ties were working on such an alternative kind of entanglement. They had discovered a correlation between the sun's activities and the price of stock on Wall Street, while their boss, Alan Greenspan, managed the most influential and powerful central bank in the world.

If you are a little open-minded, you might start to think that if some scholar at the Federal Reserve is engaged in this research, then it is just as rational to accept my Quantum Trading algorithms, presented in Chapters 5 and elsewhere, which are based on similar entanglements.

You may begin to contemplate, like Shakespeare, that "there are more things in heaven and earth...than are dreamt of in your philosophy" (*Hamlet*, act I, scene 5).

Let's take a look at the abstract from this study published by the Federal Reserve of Atlanta:

> *Explaining movements in daily stock prices is one of the most difficult tasks in modern finance. This paper contributes to the existing literature by documenting the impact of geomagnetic storms on daily stock market returns. A large body of psychological research has shown that geomagnetic storms have a profound effect on people's moods, and, in turn, people's moods have been found to be related to human behavior, judgments, and decisions about risk. An important finding of this literature is that people often attribute their feelings and emotions to the wrong source, leading to incorrect judgments. Specifically, people affected by geomagnetic storms may be more inclined to sell stocks on stormy days because they incorrectly attribute their bad mood to negative economic prospects rather than bad environmental conditions. Misattribution of mood and pessimistic choices can translate into a relatively higher demand for riskless assets, causing the price of risky assets to fall or to rise less quickly than otherwise. The authors find strong empirical support in favor of a geomagnetic storm effect in the stock returns after controlling for market seasonal, and other environmental and behavioral factors. Unusually high levels of geomagnetic activity have a negative, statistically and economically significant effect on the following week's stock returns for all U.S. stock market indices. Finally, this paper provides evidence of substantially higher returns around the world during periods of quiet geomagnetic activity.*

The authors began their study observing that:

> *While it is the geomagnetic storms (GMS) that give rise to the beautiful Northern lights, occasionally they can also pose a serious threat for commercial and military satellite operators, power companies,*

astronauts, and they can even shorten the life of oil pipelines in Alaska by increasing pipeline corrosion.

Most importantly, geomagnetic storms can pose a serious threat for human health. In Russia, as well as in other Eastern and Northern European countries, regular warnings about the intensity of geomagnetic storms have been issued for decades. More recently, the research on geomagnetic storms and their effects started to become more and more important in several other countries such as the United States, the United Kingdom, and Japan. Now, we can get regular updates on the intensity of the geomagnetic activity from the press, the Internet, and the Weather Channel."

DEFINING SUN SPOTS AND GEOMAGNETIC STORMS

Anna Krivelyova and Cesare Robotti continue their study by observing:

Geomagnetic storms are worldwide disturbances of the earth's magnetic field, distinct from regular diurnal variations. The sun continuously emits a "solar wind" (often called by specialists the solar wind plasma) in all directions. It is very fast and highly variable, both in speed and in density. This wind blows radially away from the sun and always contains a magnetic field which is also highly variable in magnitude and direction. Because the sun rotates completely around in about 27 days, as seen from the earth, the average magnetic field contained within the solar wind forms a spiral pattern. When the magnetic field direction within the solar wind is directed opposite to the earth's magnetic field, then large geomagnetic storms can occur. Specifically, the sun, from time to time, emits "bubbles" (or coronal mass ejections) which are faster, often more dense than normal and contain higher magnetic fields. These bubbles travel away from the sun at about 2 million miles per hour. If "bubbles" leave the right place on the sun to reach earth, they travel the 93-million-mile distance in about 40 hours. Coronal mass ejections occur more often when the sun is more active, and sunspots are more numerous during such times. Since sunspot activity peaks every 11 years, geomagnetic storms exhibit some cyclicality as well.

Figure 4.1 shows that geomagnetic storms correlate with sunspots.... Moreover, the sunspots and the GMS cycles are not perfectly synchronized. Physicists at the University of California, San Diego and Japan's Nagoya University have improved geomagnetic

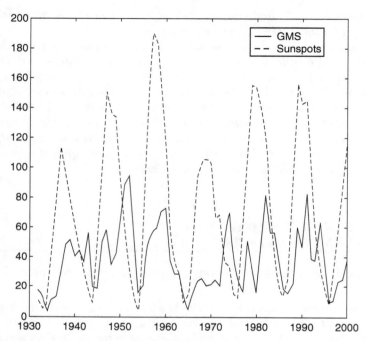

FIGURE 4.1 Geomagnetic Storms versus Sun Spots
Source: Federal Reserve of Atlanta.

storms predictions dramatically in the past few years by developing
a method of detecting and predicting the movements of these geomagnetic storms in the vast region of space between the sun and the earth.
Forecasts of geomagnetic activity at different horizons are available
from NASA and various other sources. Geomagnetic storms are clas-
sically divided into three components or phases [see, for example,
Persinger (1980)]: the sudden commencement or initial phase, the
main phase, and the recovery phase.

The figure [Figure 4.1] displays the line graph of the average
number of sunspots and geomagnetic storms (vertical axis) per year.
Geomagnetic data can be downloaded from the following web site:
ftp://ftp.ngdc.noaa.gov/STP/GEOMAGNETIC_DATA/INDICES/.

Geomagnetic storms are predictable and persist for periods of
two to four days. On average, we have 35 stormy days a year
with a higher concentration of stormy days in March–April and
September–October. Geomagnetic storms have been found to have
brief but pervasive effects on human health and have been related to
various forms of mood disorders. Geomagnetic variations have been

correlated with enhanced anxiety, sleep disturbances, altered moods, and greater incidences of psychiatric admissions [Persinger (1987, page 92)].

In a study on GMS and depression, Kay (1994) found that hospital admissions of predisposed individuals with a diagnosis of depression rose 36.2 percent during periods of high geomagnetic activity as compared with normal periods. A phase advance in the circadian rhythm of melatonin production was found to be the main cause of the higher depression rates. Raps, Stoupel, and Shimshoni (1992) document a significant 0.274 Pearson correlation between monthly numbers of first psychiatric admissions and sudden magnetic disturbances of the ionosphere. Usenko (1992) finds that, on heliomagnetic (solar) exposures, pilots with a high level of anxiety operate at a new, even more intensive homeostatic level which is accompanied by a decreased functional activity of the central nervous system. The latter leads to a sharp decline in flying skills. Kuleshova et al. (2001) document a substantial and statistically significant effect of geomagnetic storms on human health. For example, the average number of hospitalized patients with mental and cardiovascular diseases during geomagnetic storms increases approximately two times compared with quiet periods. The frequency of occurrence of myocardial infarction, angina pectoris, violation of cardial rhythm, acute violation of brain blood circulation doubles during storms compared with magnetically quiet periods. Oraevskii et al. (1998) reach similar conclusions by looking at emergency ambulance statistical data accumulated in Moscow during March 1983-October 1984. They examine diurnal numbers of urgent hospitalization of patients in connection with suicides, mental disorders, myocardial infarction, defects of cerebrum vessels, and arterial and venous diseases. Comparison of geomagnetic and medical data show that at least 75 percent of geomagnetic storms caused increase in hospitalization of patients with the above-mentioned diseases by 30–80 percent on average. Zakharov and Tyrnov (2001) document an adverse effect of solar activity not only on sick but also on healthy people: "It is commonly agreed that solar activity has adverse effects first of all on enfeebled and ill organisms. In our study we have traced that under conditions of nervous and emotional stresses (at work, in the street, and in cars) the effect may be larger for healthy people. The effect is most marked during the recovery phase of geomagnetic storms and accompanied by the inhibition of the central nervous system." Using a sample of healthy people, Stoilova and Zdravev (2000) and Shumilov, Kasatkina, and Raspopov (1998) reach similar conclusions. Tarquini, Perfetto, and Tarquini (1998) analyze the relationship between geomagnetic activity, melatonin, and seasonal depression. Specifically, geomagnetic storms, by influencing the activity of the pineal gland, cause imbalances and disruptions of the circadian rhythm of melatonin production, a factor that plays an important role in mood disturbances. Abnormal melatonin patterns have been closely linked to a variety of behavioral changes and mood disorders. In general, studies have reported decreased nocturnal melatonin levels in patients suffering from depression. An unstable circadian secretion pattern of melatonin is also associated with depression in SAD.

The relationship between melatonin, day length variation rate, and geomagnetic field fluctuations has also been analyzed by Bergiannaki, Paparrigopoulos, and Stefanis (1996). Sandyk, Anninos, and Tsagas (1991), among others, propose magneto- and light therapy as a cure for patients with winter depression: "In addition, since the environmental light and magnetic fields, which undergo diurnal and seasonal variations, influence the activity of the pineal gland, we propose that a synergistic effect of light and magnetic therapy in patients with winter depression would be more physiological and, therefore, superior to phototherapy alone." Even if geomagnetic activity is more intense during spring and fall [see Figure 4.2)], leading to increased susceptibility for desynchronization of circadian rhythms, geomagnetic storms, and their effects on human beings are not purely seasonal phenomena. This evidence complements and contrasts additional medical findings on the link between depression and SAD, a condition that affects many people only during the seasons of relatively fewer hours of daylight. While SAD is characterized by recurrent fall and winter depression, unusually high levels of geomagnetic activity seem to negatively affect people's mood intermittently all year long. Moreover, the response of human beings to a singularly intense geomagnetic storm may continue several days after the perturbation has ceased. In summary, there seems to be a direct causal relationship between geomagnetic storms, and common psychological disorders and geomagnetic activity seems to affect people's health with a lag.

The figure displays the bar graphs of the returns on the NASDAQ, S&P 500, AMEX, and NYSE (NY) stock market indices during normal days (left column) and bad days (right column). We define the six calendar days after a storm as bad days and the remaining calendar days as normal days.

Experimental research in psychology has documented a direct link between mood disorders and decision making. Hirshleifer and Shumway (2003) provide a detailed summary of these studies.

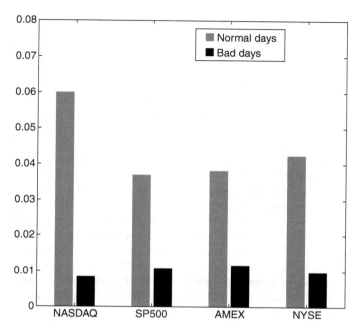

FIGURE 4.2 U.S. Stock Returns during Normal Days and Bad Days
Source: Federal Reserve of Atlanta.

For example, Wright and Bower (1992) show that, when people are in bad moods, there is a clear tendency for more pessimistic choices and judgments. Mood mainly affects relatively abstract judgments, about which people lack concrete information. Bad moods also lead to a more detailed and more critical analytical activity [Schwarz (1986), Petty, Gleicher, and Baker (1991)]. Loewenstein (2000) discusses the role of emotions in economic behavior, Johnson and Tversky (1983) find that mood has strong effects on judgments of risk. Frijda (1988), Schwarz (1986), Clore and Parrott (1991), Clore, Schwarz, and Conway (1994), and Wilson and Chooler (1991), among others, show that emotions and moods provide information, perhaps unconsciously, to individuals about the environment. An important finding of this literature is that people often attribute their feelings and emotions to the wrong source, leading to incorrect judgments. Specifically, people affected by GMS may be more inclined to sell stocks on stormy days, by incorrectly attributing their bad mood to negative economic prospects rather than bad environmental conditions. Market participants directly affected by GMS can influence overall market returns according to the principle that market equilibrium occurs at prices where marginal buyers are willing to exchange with marginal sellers. Misattribution of mood and pessimistic choices can translate into a relatively higher demand for riskless assets, causing the price of risky assets to fall or to rise less quickly than otherwise. Hence, we anticipate a negative causal relationship between patterns in geomagnetic activity and stock market returns. Medical findings do not allow us to identify a precise lag structure linking geomagnetic storms to psychological disorders, but make it clear that the effects of unusually high levels of geomagnetic activity are more pronounced during the recovery phase of the storms [see, for example, Zakharov and Tyrnov (2001), Halberg et al. (2000), and Belisheva et al. (1995)]. Hence, we use daily data to empirically investigate the link between stock market returns at time t *and GMS indicators at time* t−k, *with choice of* k *motivated below. Therefore, against the null hypothesis that there is no effect of GMS on stock returns, our alternative hypothesis is that psychological disorders brought on by GMS lead to relatively lower returns the days following intense levels of geomagnetic activity. Notice that the relation between GMS and the stock market is not subject to the criticism of datasnooping. Exploration of whether this pattern exists was stimulated by the psychological hypothesis, and the hypothesis was not selected to match a known pattern."*

PROPOSED TRADING STRATEGIES

In the same study, A. Krivelyova and C. Robotti proposed their strategy to take advantages of the correlation. They pointed out:

Figures 4.3 to 4.4 show that returns during "normal" days are substantially higher than returns on "bad" days for most of the stock market indices in our sample. A natural question related to this empirical finding is whether we can use the information displayed in Figures 4.3 to 4.4 to build exploitable trading strategies. In forming simple trading strategies based on the GMS effect, we face transaction costs as the main problem. Even though geomagnetic storms are predictable, their frequency, intensity, and persistence varies over time. Shortening the calendar window that we use to define "bad" days would help us to pinpoint the days characterized by particularly low (and often negative) returns, but would significantly increase the number of transactions that we have to make.

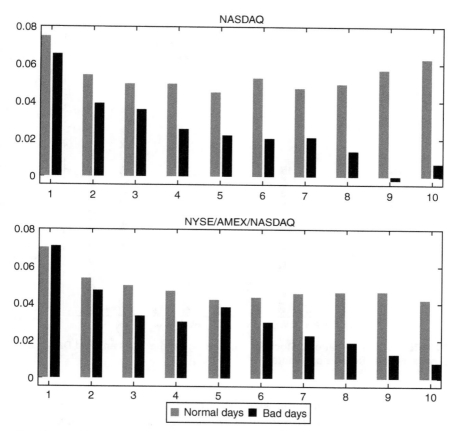

FIGURE 4.3 Returns during Normal Days and Bad Days for U.S. Size Deciles
Source: Federal Reserve of Atlanta.

The figure displays the bar graphs of the returns on the NASDAQ and NYSE/AMEX/NASDAQ size deciles during normal days (left column) and bad days (right column). We define the six calendar days following a geomagnetic storm as bad days. We define the remaining calendar days as normal days. Large cap = 1, ..., Micro Cap = 10.

The figure displays the bar graphs of the returns on the World, Canadian (CAN), Swedish (SWE), British (UK), Japanese (JAP), Australian (AUS), New Zealander (NZ), South African (SA), and German (GER) stock market indices during normal days (left column) and bad days (right column). We define the six calendar days after a storm as bad days and the remaining calendar days as normal days.

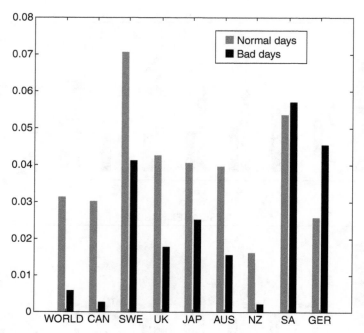

FIGURE 4.4 International Stock Returns during Normal Days and Bad Days
Source: Federal Reserve of Atlanta.

*One simple trading strategy based on our six day calendar
window described above would be the following. An individual might
try to hold the world market portfolio during "normal" days and
switch his investments toward safer assets such as the 3-month Eurodollar deposits during "bad" days. This trading strategy would require rebalancing the GMS-based portfolio on average 26 times a
year. Ignoring transaction costs, this trading rule would generate
an average annual return of 7.5 percent, while a buy and hold policy
would yield a 6.4 percent annual return. The GMS-based portfolio
would also deliver a standard deviation which is 14 percent lower
than the standard deviation of the benchmark portfolio. However,
no individual investor can ignore transaction costs. By referring to
Huang and Stoll (1997), Hirshleifer and Shumway (2003) approxi-
mate transaction costs with the cost of trading one S&P 500 futures
contract as a fraction of the contract's value and come up with an
estimate. The 3-month Eurodollar deposit rate is from the Board of
Governors of the Federal Reserve System. The series spans the entire length of the return on the world market portfolio. Berkowitz*

et al. (1988) estimate the cost of a transaction on the NYSE to be 0.23 percent. One of the largest institutional investors worldwide, the Rebecco Group, estimates transaction costs in France 0.3 percent, Germany 0.5 percent, Italy, 0.4 percent, Japan 0.3 percent, the Netherlands 0.3 percent, and the United States 0.25 percent. In the UK, the costs of a buy or sell transaction are 0.75 percent or 0.25 percent, respectively. Solnik (1993) estimates round-trip transaction costs of 0.1 percent on future contracts. 22 of one basis point per transaction. With costs of two basis points roundtrip, our GMS strategy would generate an average annual return of 7.25 percent, while the buy and hold policy would always yield a 6.4 percent annual return. The breakeven point is represented by eight basis points roundtrip. In this latter case, the GMS-based strategy and the buy and hold strategy would deliver almost identical annual returns. Even if our GMS-based strategy seems to produce small trading gains, an individual could increase the expected return to his investments by altering the timing of trades which would have been made anyway—executing stock purchases scheduled for "normal" days on "bad" days and delaying stock sales planned for "bad" days on "normal" days. There might be more effective ways of taking advantage of the GMS effect in stock returns. One possibility would be to use derivative securities as a hedging device. Trading against incoming storms by buying put options on stock market indices might turn out to be a valid strategy.

You could trade the U.S. stock indexes tapping into this strategy if you follow sun-spot activity with different scientific reports issued by international astronomical agencies. The problem is that sun-spot activity, and in turn the geomagnetic storms it precipitates, are not predictable far in advance, even if you can know the time when the event occurs.

Because of little advance notice, it seems difficult to create a real trading system based only on geomagnetic storms. Furthermore, even though you would trade the major index using exchange-traded funds (ETFs) or futures, this method doesn't provide any price level indication to open and close a trade. This is a very large handicap. So even if there is a relevant correlation between these two phenomena, it's not easy making money this way.

Fortunately our Quantum Trading algorithms allow us to draw on the charts far in advance the most probable price and time of reversal zones where we expect a major or intermediate top or bottom in stocks, commodities, and currency markets.

CHAPTER 5

P-Space Structure and Quantum Trading Algorithms

I n the previous chapters we introduce the main concepts used to develop the Quantum Trading system based on QPLs. We offer a taste of how QPLs work in several charts. We also review the basic concepts of Einstein's theory of relativity, demonstrating how space is not uniform and how gravity, generated by the presence of a celestial body's mass, curves space.

One effect of curved space is that photons, or light particles, deviate from their linear path. The concept of a curved space and the phenomena of light deflection are fundamental for understanding P-Space and QPLs.

Now the moment has arrived to finally understand how to build the P-Space and study the equations to calculate QPLs. Then we'll know how to draw QPLs on our favorite stock or currency chart.

Let's briefly review the main properties of P-Space. P-Space is a multidimensional, virtual space composed of securities prices, time, and celestial objects within our solar system in motion, which curve space-time due to their mass, or gravitational effect. Other properties of P-Space include:

- It is an interactive, virtual structure ruled by entanglement and nonlocality.
- The price movement of a stock or currency, and its top and bottom in P-Space show movements similar to a particle of light and an electron.
- P-Space is ruled by laws similar to the theory of relativity and quantum physics.

A study of geomagnetic storms, published by the Federal Reserve of Atlanta and discussed in Chapter 4, proposes that geomagnetic storms are caused by sun-spot activity. The sun is the protagonist of a correlation between its spot activity and stock market returns. If we focus on this concept we can also consider the activity or movement of other celestial bodies in our solar system, such as the major planets, to discover other correlations.

If you place the planets in our P-Space you discover that, using a special operator to measure its curvature, their mass can curve the space-price-time enough to deflect the movement of the price of a security, which we treat as a photon in our example.

The effect of the curvature in the P-Space then is amplified. In our solar system the sun has the largest mass to curve the space and produces a higher light deflection effect.

However, in our P-space, the other planets can significantly curve the space in which the price moves, especially those with a larger mass. In P-Space the deflection magnitude is so high that the price can reverse its path. In our daily lives we perceive these phenomena as bottoms and tops in a security chart.

Basically the P-Space is a separate, virtual universe ruled by the theory of relativity and quantum laws where entities such as price exist, move, and act according to these laws.

To calculate the points of our P-space and find the higher level of curvature, we need a special operator. We call that operator the "P-Space Operator" (PSO).

To understand the importance of the PSO, it is useful to remember that Einstein didn't make any progress on his theory for more than 10 years, until he met Riemann and learned how to use Riemann's curvature tensor in his calculation of relativity. It allowed Einstein to calculate the magnitude of light deflection. The principle of deflection is shown in Figure 1.5 in Chapter 1.

PSO represents in my Quantum Trading model what Riemann's curvature tensor represented for Einstein in developing his theory of relativity. From now on we'll speak of the deflection of price instead of light deflection phenomena.

Einstein finally received official confirmation of general relativity when Arthur Eddington observed a solar eclipse on Principe Island in New Guinea in 1919. Fortunately, we do not need to wait for the next solar eclipse, as Einstein did, to confirm Quantum Trading theory. We can just apply Kepler's three laws for spotting the position of a planet and calculating its mass for a given period to measure its effects on price behavior.

If the price of a stock or currency is deflected in P-Space, it means that it has been affected by the mass of a planet and we'll see a top or a bottom, or at least a correction in case of minor curvatures for the corresponding price.

In geographical astronomy the geocentric position of a planet on the ecliptic is individuated by a set of spatial coordinates such as longitude, right ascension, declination, and latitude. You can also use heliocentric co-ordinates to spot the position of a planet. In this case you are not observing everything from the Earth, as in the case of geocentric coordinates, but rather your point of observation changes.

A longitude's number always ranges between 0 and 360 degrees because planets, according Kepler's three laws, move on an elliptical plane. In our P-space we commonly use both geocentric and heliocentric coordinates to spot curvature points that are then transformed into prices. From this point on we'll use the term *geo* to refer to geocentric and *helio* for heliocentric.

The ecliptic is the plane of the Earth's orbit around the Sun. The Earth's orbit, viewed from the side, like a circle viewed from the side, is a plane. So the ecliptic is the plane of Earth's circular orbit extended to infinity.

Ecliptic describes the centerline of what ancient Greek astronomers called the Zodiac, which extends eight degrees above and below the ecliptic. In other words, the Zodiac is a belt 16 degrees wide centered on the ecliptic, and it is the track on which planets move.

Now we will present Johannes Kepler's three laws, which help us to calculate the position of a planet.

Kepler's Three Laws

Johannes Kepler developed three laws that describe the motion of the planets across the sky.

1. The law of orbits: All planets move in elliptical orbits, with the sun at one focus.

2. The law of areas: A line that connects a planet to the sun sweeps out equal areas in equal times.

3. The law of periods: The square of the period of any planet is proportional to the cube of the semimajor axis of its orbit.

We could dig into all of Kepler's equations and the calculations you need to fix the position of the planets in P-Space and then calculate the higher-level curvature point; however, this would be superfluous since you can find them in many books of physics and astronomy. You can find free software on the Web, or you can consult the NASA ephemeris that shows the position of the planets in our solar system day by day.

Now we'll finally explain how to use the PSO to convert curvature points into prices to find unconventional, but very effective resistance and support levels.

This happens for the price deflection phenomenon due to the presence of a mass in the P-Space. Basically, in P-Space there are 360 possible positions around the ecliptic. When a planet occupies one of these positions, it curves the P-Space in the same way that a heavy steel ball in the middle of a taut tablecloth will create an indentation, as we discuss in Chapter 1.

In P-Space, a given position of a planet can be transformed into a given price for a given time by making use of the PSO.

PRICE SPACE OPERATOR (PSO)

The PSO indicates the price of a security corresponding to a certain degree of curvature of P-Space. The PSO identifies powerful points of Support and Resistance.

CALCULATING A QPL

Tapping into the PSO, each day I can get a given potential price related to the points where the highest levels of curvature of the P-Space occur for a certain planetary mass. So, I obtain different price-points for each day of the year.

Below you will find the formula to calculate the PSO and PSO harmonics:

$$\text{PSO harmonics} = (N \times 360°) + \text{PSO}$$

or

$$(1/N \times 360°) + \text{PSO}$$

where   PSO = Longitude × conversion scale
      Long = Planet longitude (geo or helio)
       CS = Conversion scale
       N = The harmonic you need to use to approach the nearest relevant QPL of a security.

PSO Harmonics

When you are dealing with price securities showing higher prices than 360 degrees, you have to use the N harmonic to get a value closer to the securities actual price.

If you are dealing with price of a currency where the price is near 1, you have to use the inverse harmonic of $1/N$.

Assuming that CS is equal to 1 then PSO would be equal to Longitude.

If you want to calculate the curvature of P-Space and its price-point generated by the planet Saturn helio on October 11, 2007, the day of the historical S&P 500 top, you first need to find the helio longitude of Saturn on that day. This is equal to 150°.33.

$$PSO = Long \times CS$$

So, using CS = 1, PSO = 150.33.

S&P 500 shows a price of 1586, which is higher than 360, so we need to use PSO harmonics; otherwise, we would only have a price in the range of 0–360 degrees.

We'll add 360 to 150.33 n times.

The P-Space generates curvature points simultaneously at the following prices:

150.33	
150.33 + 360	510.33
150.33 + (360 × 2)	870.33
150.33 + (360 × 3)	1230.33
150.33 + (360 × 4)	1590.33
150.33 + (360 × 5)	1950.33
And so on.	

Because the S&P 500 futures price is 1586, very close to 1590.03, which is the fifth PSO harmonic, I anticipated a reversal on October 11, 2007. The historical top was made on exactly that day.

PSO EQUALS A QPL's VALUE ON A CERTAIN DAY

The PSO, or its harmonics, show the price level where you can find a QPL on the same day.

Usually, in the case of swing trading where we want to trade the market for a movement lasting at least three to four days, we consider a lost motion, or a tolerance, of seven points with respect to a QPL.

In this case, at 1583—seven points below the QPLs passing at 1500.03—and following our rule, we opened a short position on S&P 500 selling futures.

The target price to take profit is provided by the lower QPL passing at 1230.33 on the same day. It changes its value every day. In this case we are dealing with a slow planet, and the price associated with it in P-Space changes only a few cents a day, assuming that N is equal to 1.

In this example, we can say that within a few weeks the target price will be basically the same with respect to the one that we have calculated on October 11, 2007.

In fact, if we want to precisely calculate the same QPL value on October 22, 2007, the day when price reached the target price, we need to first check that Saturn's helio longitude is $150°.72$, very close to the initial value of October 11. For instance, after 6 months, on April 22, 2008, the helio longitude is $158°.19$.

You can draw on an S&P 500 chart these PSO points, and if you join them together you will obtain a line, which is the QPL.

A QPL, as you remember from the previous chapters, is generated by the curvature activity of an object in our solar system within the P-Space. If a price happens to be very near or touch a QPL, it can be deflected from its original trend and we'll see a reversal in a stock, commodity, or currency chart.

OBTAINING QPL

A QPL is obtained by joining day-by-day points of the PSO or the same PSO harmonic points.

This means that we have to exclusively join the points belonging to the second PSO harmonic to the third PSO harmonic, to the fourth PSO harmonic, and so on, without mixing one harmonic point with another.

To summarize: first we need to know the position of the planets, then we bring them into the P-Space. We then use a simple operator to transform the planet's longitudes into prices where the curvature is at its maximum level and could cause a reversal. We calculate for each day of the year the position in P-Space—expressed in terms of price—of all potential curvature points. Each individual point is a component of a QPL. If we join all of the points we get a support or resistance line that is a full QPL.

In case the price is higher than 360, we join all of the points of the same PSO N harmonic to get a full QPL. Why do we choose, among so many QPLs, only the fourth harmonic at 1590.33 for our October 11, 2007, S&P 500 trade? Simply because the fourth harmonic showed, on that day, a price that was the closest to the real S&P 500 price.

When the mass of two celestial bodies occupies the same degree of longitude of the ecliptic, the curvature point is much bigger than usual. The S&P 500 trade that I discuss with my friends in Chapter 1 is based on this model. Two planets occupied the same degree of the ecliptic;

astronomically this is called a conjunction. Their masses were added together and curved the space, and the PSO helped us to calculate their re-
spective QPLs. In the proximity of the conjunction, the two QPLs stemming
from each planet crossed and the price of the S&P 500 was tangent. So we
faced a big drop of the index, as you can see in Figure 1.1 in Chapter 1,
because price and the two QPLs were simultaneously in the same spot of
the P-Space.

While it is easy to understand what "longitude" is, we need to discuss
the conversion scale more extensively.

The conversion scale is an operator using a quantum approach for
the solution of the curvature magnitude found in certain price points of
P-Space.

Max Planck was the one who first took a quantum leap in particle
physics. Planck stated that the problem of the emission of varying degrees
of ultraviolet could be solved. The electrons are assumed to emit or absorb
energy only in certain specific, discontinuously discrete amounts—which
he called "quanta" of energy.

Also in our P-Space, the curvature magnitude (calculated as PSO =
Longitude × CS) depends on a factor increasing or decreasing its value by
discontinuously discrete amounts, or "quanta." This is mandatory for the
calculation of curvature of our P-Space.

It seems that Gann was also interested in finding a relationship between Kepler's three laws and stock prices. However, there is no evidence
that he used a quantum approach for his calculations in this field; instead,
he remains quite general and vague on this subject.

In fact, the value that can be assumed by CS in P-Space can be only
one of the following numbers:

1, 2, 4, 8, 16, 32, and so forth. That is equal to 2 to the n power. Or,
$1/2$, $1/4$, $1/8$, $1/16$, $1/32$, and so forth. That is equal to 1 divided by 2 to the
n power.

This means that the magnitude curvature in P-Space can be increased
or decreased only in a "quantum" way.

CS is the suboperator we use to find different prices generated by
P-Space.

If we have an index like the DAX at 8,000 or the Dow Jones at 11,000,
we need to use a bigger value for CS to get a better fit and more accurate
information. But try to remember that this value can be selected only from
between the two equations described above.

For example, Uranus on October 11, 2007, showed a helio-longitude
equal to 347.04. If we want to check out where the PSO is located using a
CS value equal to 8, we will obtain the following results:

$$PSO = Long \times CS$$
$$PSO = 347.04 \times 8 = 2,779.20$$

FIGURE 5.1   Dow Jones CBOT Future Chart: Historical Top and QPL

If we add CS 8 × 360 = 2880 to 2,779.20, we get the second harmonic at 5659.20. If we add three times 2880 to 5659.20, we get another harmonic at 14,299.20, which is very close to the value of the historical top of the CBOT Dow Jones future at 14,270 on October 11, 2007, as you can see in Figure 5.1 at point A.

Computer processors are based on binary code and the development of computer power can be expressed in a quantum way, following the two of the binary pair to the n power. You always have computers with 8, 16, 32, 64, and 128 bits, and you will never see a computer with 17, 23, or 35 bits.

Our PSO works using the same scale. If you have software that helps you to automatically display all QPLs on your favorite chart, it's very easy to get all of these calculations within a few seconds.

CALCULATING A QPL's SUBHARMONICS

A QPL's subharmonic (QPLSH) allows us to spot resistance and support level prices where we can find intermediate tops and bottoms. In fact, besides using QPLs to forecast historical tops and major bottoms, it is very useful to buy and sell on price levels that we can trade more frequently.

QPLSHs are one of the most important tools in Quantum Trading. So, let's see how to calculate them.

We start by taking the figure of the circle. If we take the circle as a unit, this corresponds to QPLs and their main harmonics. The circle can also be divided in the following ways:

- If we divide the circle, which is composed of 360 degrees, by two, we obtain 180 degrees. It corresponds both to 180 degrees, or a 50 percent division of a range.
- If we divide it by four, we obtain 90 degrees, which is equal to a 25 percent division of the range.
- If we divide it by eight, we obtain 45 degrees, which is equal to a 12.5 percent division of the range.
- We can also divide the circle by three, and we will obtain 120 degrees.
- Dividing the circle by six we obtain 60 degrees.

If you want, you can continue dividing the circle further.

We can use the number we obtained dividing the circle to calculate the QPLSHs, which are a division of two subsequent QPLs. Recalling the previous calculation of helio-Saturn QPL on October 11, 2007, and starting from Long $= 150°.33$, we have the following for a CS $= 1$.

The P-Space generates curvature points simultaneously at the following prices:

$$
\begin{array}{ll}
150.33 & \\
150.33 + 360 & 510.33 \\
150.33 + (360 \times 2) & 870.33 \\
150.33 + (360 \times 3) & 1230.33 \\
150.33 + (360 \times 4) & 1590.33 \\
150.33 + (360 \times 5) & 1950.33 \\
\text{And so forth.} &
\end{array}
$$

So, we can take the two QPLs passing at 1590.33 and 1230.33 and calculate the harmonics adding CS $(=1) \times 180 + 1590.33$, and we arrive at 1770.33.

The 180 QPL passes at 1770.33, or at 1410.33 (1230.33 + 180).

If you want to calculate a 90-degree QPLSH, you proceed in the same way, adding 90 instead of 180 to a QPL. We take 1230.33 QPL and we simply add 90, obtaining 1320.33. We can obtain another 90 QPLSH passing at 1410.33, if we add 90 to 1320.33.

You can obtain all the QPLSHs you need to spot intermediate tops and bottoms, preceding as we have just explained above. Furthermore, you can also use different numbers that are further divisions of 360, dividing by two the number we used above.

EXAMPLES OF POWERFUL QPLs AND QPLSHs

In the next few pages you will find many examples of very popular stocks, stock indexes, and currencies. If you have traded these securities you will find them very interesting and understand why the price reversed at the levels where you can find both QPLs and QPLSHs.

Figure 5.2 covers the period between August 2008 and October 2009. Each letter represents the point at which the price met with a support or resistance QPL. At point A we can see a top and at point B a bottom that in a while will became a double bottom. At point C you find another intermediate top.

The currency chart in Figure 5.3 shows the performance of the FX spot GBP–USD from May 2009 until August 2010. The five points (A, B, C, D, E) identify tops and bottoms that corresponded with QPLs.

In Figure 5.4, we can see how QPLs channel the price until the price bounces off a support and leaps up to point C, where it finds a resistance.

Figure 5.5 shows a major top at point B, stemming from the bottom of the QPL you find at point A. The price is channeled until point B, where a QPL acts as a strong resistance. Please notice how point B points out the first of a series of tops lying on this same price level. Finally, the price reverses until it reaches the next QPL at point C.

We can see in Figure 5.6 that price always reacts after touching QPLs. Point A marks a resistance that causes the price to fall until point B, a support. The price then breaks through the middle QPL touching point C, where we have another reversal and find support at point D.

FIGURE 5.2 USD-Yen Currency Chart: Support and Resistance QPLs

FIGURE 5.3 GBP-USD Support and Resistance QPLs

Figure 5.7 shows the major top at point A and the major bottom at point B.

NOT ALL QPLs ARE ALIKE

After having learned how to calculate QPLs and QPLHs, you are ready to learn more about the nature of different QPLs.

As we mentioned before, a QPL is generated by the PSO, which indicates in turn the price of a security corresponding to a certain curvature in the P-Space. The curvature is generated in turn by the presence of a mass

FIGURE 5.4 USD–CAD FX Spot Chart and QPL

FIGURE 5.5 AUD-USD FX Spot Chart and QPL

in P-Space associated with a planet. We showed that the curvature generated by Saturn provided through the PSO the historical top of the S&P 500 at 1583.

Not all of the QPLs are equally strong. Some are able to redirect the price from a bull or a bear campaign lasting years, and they show historical or major tops and bottoms. Others are weaker and the degree of P-Space curvature they convey can create only minor reversals. From now on we'll name the different QPLs with a letter because otherwise someone might

FIGURE 5.6 AUD-USD FX Spot Chart and QPL (Longer Period)

FIGURE 5.7   AUD-USD FX Chart and Geo QPL: Major Top and Bottom

think that the planets themselves are able to cause a reversal in the financial markets and could be inclined to incorrectly associate the celestial object of our P-Space with astrology.

This improper association would be absolutely wrong because P-Space is a virtual space simulating Einstein's space-time, and PSO is a mathematical operator using increasing or decreasing factors, exactly like discrete amounts of "quanta" are used in quantum physics. Furthermore, PSO has nothing to do with the nature of planets studied in astrology, being instead a tool to calculate the curvature of P-Space, exactly as Reimann's curvature tensor was used by Einstein to calculate the light deflection phenomena.

In our P-Space, instead, we speak of the phenomena of price deflection.

So, from now on the letters we associate with the celestial objects in our P-Space are the following:

P = Pluto, N = Neptune, U = Uranus, S = Saturn, J = Jupiter, M = Mars, Su = Sun, V = Venus, Me = Mercury.

QPL Hierarchy

P, N, U, S, and J QPLs show the major tops and bottoms.
M QPLs show intermediate tops and bottoms.
Su, V, and Me QPLs show the minor reversals.

When you calculate the curvature of P-Space to forecast a major reversal, do not use Su, V, or Me QPLs alone, because you will be misled.

TRADING WITH INTRADAY QPLs

In the previous sections we have seen many examples of day charts, just to get acquainted with QPLs, but don't think that they can apply only on a daily chart and can be used only for long-term trading.

QPLs can also be used for intraday trading, and you can apply them to any time-frame chart, such as 5-, 15-, 30-, or 60-minute charts.

In the following charts you will see many examples of this.

In Figure 5.8 we have Hel M QPLSH (60-minute) and CS: 0.0625.

Notice in Figure 5.9 how the QPLSH eventually corresponds to a classical trend line. Actually a trend line needs at least two points to be drawn, while QPLs and QPLSHs do not need any points to be found, because they are generated by P-Space equations.

When Paul Dirac was awarded the Nobel Prize in 1933, he said in his acceptance speech, "The method of theoretical physics should be applicable to all those branches of thought in which the essential features are expressible with numbers."

He praised the importance of interdisciplinary study. The quantum world and the financial market actually have some characteristics in common. I hope that this study can represent how the appearance of randomness and mathematical similarity can function as a reservoir of ideas and creativity for both the quantum and financial world.

In the next chapters we will review W. D. Gann's life and trading ideas.

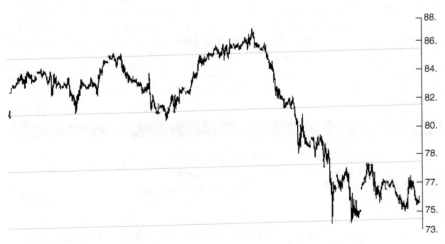

FIGURE 5.8 Crude Oil Future 30-Minute Chart and Intraday QPL (CS = 0.0625)

FIGURE 5.9 S&P 500 Index 60-Minute Intraday Chart, Hel V QPLSH (180) (CS = 0.25)

In particular, we'll study how Gann Angles really work because, despite many books about his techniques, this subject has never been treated in an extensive way. So, even with several kinds of software offering Gann Angles within their arsenals of tools, many people disregard the angles because they do not know how to use them, and they do not offer a rational approach for the conversion scale that relates time and price units.

We'll propose a quantum use of Gann Angles. Then we will use QPLs and Gann Angles simultaneously, and by doing so, we'll obtain powerful entry-levels for very profitable trades.

CHAPTER 6

The Life and Contributions of W. D. Gann

A Forerunner of Quantum Trading

Sometimes events occur that change our lives forever, and that's exactly what happened to young William Delbert Gann. One evening in Lufkin, Texas, his dad came home with some unexpected gifts.

Mr. Gann entered with a jute bag in his hand, stopped at the entrance and said to his sons, "Boys, I have a present for you." He tipped the bag upside down and shiny new shoes came flying out, scattering onto the floor. His ten sons yelled with joy and threw themselves onto the treasure, trying on different pairs to find the right fit. For months, the boys had been walking to school in ragged, worn-out shoes. At the time, a good pair of shoes was not only a necessity, but also a sign of safety and security.

That night the Gann family celebrated; the father had brought home the most delicious foods for the family as well as new dress fabric for his wife, who was the mother of the 10 young'uns running around the house.

The father felt proud of his achievements. After enduring much poverty and hardship, his family was finally able to enjoy some moments of joy. This night would always be remembered.

But while everyone was singing and dancing to the tunes played on a fiddle, the eldest son, 11-year-old W. D., sat in the corner by himself.

"What's wrong?" asked his father, "Why aren't you celebrating with the others?"

"It's just that . . ."

"Come on, spit it out."

"I was wondering why you were able to bring these shoes home tonight. We've been asking for new shoes for months. You knew our old ones were hurting our feet and you always told tell us that you couldn't afford them because we were already in debt. What happened tonight?"

"Well, let's just say we got really lucky today. I sold the harvest at a much higher price than last year. The price climbed significantly in the past few days and our profits almost doubled. Also, on the commodities exchange the price of cotton surged. With this sudden bounce I bought what you've been asking me to get, and what you needed most," Gann's father explained.

What we needed . . ., thought a young W. D.

He had already heard about the harvest and the wholesalers who purchased the cotton to sell it back onto the biggest commodity markets in the country. People could make huge profits by taking advantage of large price fluctuations by buying low and selling high in the same day.

The commodity exchange! Young Gann was not yet 12 but had already heard about it. His new pair of shoes came from the fluctuation of cotton prices.

"Dad, if you knew when the cotton price would climb and then waited a few days to sell, you could become very rich, right? How can you know when the markets will rise and do it again?" asked an intrigued W. D.

The father looked into his son's eyes, caressed his hair, and replied, "Son, this is a mystery that only God has the key to. Only him."

But his son, not discouraged in the least, had already made up his mind. He left the house for a bit of fresh air, lay down on the ground, and looked up at the sky. Watching a star rising above the horizon, he rested his head against a cushion of grass and said to himself, "When I grow up, I'll unlock the secrets of the commodities market. I will become rich and buy all the shoes I want. Everyone will look up to me and respect me. I will buy cotton at the lowest prices and sell at the highest ones, and this is how I'll become the richest man in Lufkin."

Young W. D. certainly grew to be one of the richest men on Wall Street. Later in his life he would frequently recall that evening, and the gift of new shoes. Gann ended up buying many pairs of nice shoes in his lifetime. He even purchased a private plane and piloted it himself! Sometimes while flying he looked up at the starry sky, thinking that only a few people actually had the courage to go out into the world and make their own destiny.

AN UNCANNY MASTER OF TRADING INTERESTED IN ATOMS AND ELECTRONS

On the morning of November 1, 1909, W. D. Gann was ecstatic because he had just successfully completed one of the most important and exciting challenges in his career as a trader. Journalists from *The Ticker and Investment Digest* had asked him to certify his performance for one entire month and see what happened. During the month of October he made 286 transactions in only 25 trading days, and he doubled his initial capital 10 times.

All of the trades were carried out by Gann under the supervision of two journalists who shadowed him during the entire period.

During those 25 trading days—at that time the stock exchange was open on Saturday, or six days per week—Gann correctly identified most of the intraday tops and bottoms of the stocks he was trading.

Furthermore, in 1908, he discovered the "market time factor," which became part of his "law of vibration." To test his new strategy, he opened one account with $300 and one with $150. It turned out to be wildly successful: Gann was able to make $25,000 profit with his $300 account in only three months; meanwhile, he made $12,000 profit with his $150 account in only 30 days! After his results were verified, he became famous on Wall Street as one of the best forecasters of all time.

But Gann did not limit himself to intraday trading. In fact, his forecasts for the long-term were uncannily accurate. He predicted far in advance the big bull campaign from 1926 to 1929. He advised his clients to take profits at the end of the first part of 1929 because he expected one of the biggest collapses of the stock market in the past 60 years.

You might ask, how did he get started? What was his secret? Was his success due to luck or bravery, or was there actually more to it? Was it perhaps his use of his famous trading tool, called "Gann Angles," or the square of nine, or the Exagon chart?

Gann gave us the solution in one of the few interviews in which he speaks of his great discovery: the "Vibration Law." "Through my method," affirmed Gann, "I can determine the price variation of every share, and, taking into consideration some temporary values, I can in most cases say how a share will behave in particular conditions."

"Stocks are like electrons, atoms, and molecules that obey with continuity to their specific individuality in giving an answer to the vibration law. If we wish to be successful in our stock market speculations and avoid losses, we have to work with the causes that form the market prices. Everything in the world is based on exact mathematical proportions, and everything is but a point of mathematical force."

What is Gann's "Vibration Law"? Unfortunately the only information we have is contained in his book, *The Tunnel Thru the Air* (pages 75–76):

> *My calculations are based on the cycle theory and on mathematical sequences. History repeats itself. That is what I have always contended—that in order to know and predict the future of*

*anything you only have to look up what has happened in the past
and get a correct base or starting point." He continues, quoting Eccl.
1.9: "'The thing that hath been, it is that which shall be; and that
which is done, is that which shall be done: and there is no new thing
under the sun...'. This makes it plain that everything works according to past cycles, and that history repeats itself in the lives of men,
nations, and the stock market."*

Fritjof Capra, author of *The Tao of Physics*, would have appreciated
this interdisciplinary approach some decades later, even if others might
consider it a mere mixed bag of ideas, ranging from the wisdom of King
Solomon, to mathematical sequences, and even to Giambattista Vico's *ante
litteram* constructivist epistemology.

Nonetheless, although it is said that Gann earned over 50 million dollars during 45 years on Wall Street, he was not only a man who accumulated
riches: He was also a big dreamer who was passionate about knowledge,
and who turned his research into one of the most successful stories in trading. He died in 1955.

Gann successfully designed a consistent and articulated strategy for
trading and profiting steadily from the stock market. He created a trading
methodology based on philosophies that were both ancient and, paradoxi-
cally, also very modern.

Studying Gann, his life, and his approach to financial markets is like
jumping down a rabbit hole. It's impossible to know what lies or waits on
the other side. Your sense of reality can be thrown off balance because you
discover that his thinking unites all sciences and philosophies in a singular,
extremely powerful weltanschauung. Gann's system is a paradox of episte-
mological realities whose paradigms are those of an entelechy of the mind
in its continual becoming.

Gann was familiar with ancient speculations about numbers, their
Pythagorean properties, and the magic of the whole and the parts, commonly described today by fractal geometry and holographic models. He
was a pioneer even in this field. He applied numerical sequences to stock
market prices, or rather numerical paradigms, to accurately interpret, anticipate, and forecast price trends, from the smallest to the biggest, according to the "analogy principle." And this was going on at the same time
that Carl Gustav Jung was studying the relationship between coemerging
structures and arguing what epistemology would have probably preferred
to negate: the principle of synchronicity. What would science be without
philosophy?

This is why we need not only a philosophy of science, but also a Gnose-
ologic epistemology.

According to Gann, everything that moves in time and space complies with the laws of physics and mathematics present in nature. He echoed Maxwell's philosophy, arguing that everything exists, but that "there is nothing in the universe but mathematical points of force."

At the same time as studying and developing his theories, Gann entered the Wall Street pit, buying, selling, and achieving stratospheric returns. In fact, during the course of a day, Gann was able to forecast the majority of the tops and bottoms of the stocks that he traded.

The press, traders, and everyone else began talking about him, fascinated by this mysterious man. In the meantime, Gann had made a lot of important friends, including the Rockefellers, John Livermore, and an American president who sought his advice. He was an intellectual giant among dwarves. Nevertheless he had one fear: to be ridiculed when discussing his interdisciplinary findings on the links that tied the sciences together, because Wall Street at that time was not interested in physics models.

Gann was a pontifex (a Latin word for "bridge-maker" and also the name of a high priest of ancient Rome) who devotedly built bridges between the known and unknown; however, he feared the anger of secular ignorance, not so much from a religious point of view, but more from an initiatic one.

Gann greatly enjoyed the music of Mozart, a freemason like himself. Many things may strike you while listening to Mozart's overtures, but one may particularly capture you: Mozart's capacity to write sophisticated music that touches the soul, dissolving a grave cadence with a scherzo, a prelude with a playful modulation.

Gann played with the finances as Mozart played with musical notes, disassembling and reassembling several trading tools that he invented to decipher the financial markets' behavior. While others were asking him about his success on Wall Street and the key to his method of trading, Gann was amazed that those around him did not understand, or even intuit, that the price of a share behaves like a particle, moving in the space-time continuum. He believed that to forecast a stock's trend it was possible to apply the same laws of physics as those used to explain gravity, motion and inertia, and the properties of particles.

For Gann, comparing the price of a stock or commodity to a particle in motion was a perfectly normal thing to do. Thanks to his physics-oriented approach he was able to earn millions in the stock market by making reliable forecasts, even in the long term.

One of his most famous forecasts was the great climb of the stock market between 1926 and 1929, as well as the crash at the beginning of 1929. Precipitated by one of the biggest speculative bubbles in the course of the preceding 60 years, the stock market crash was unexpected and severe.

This event was foreseen by Gann one year in advance. While everyone was losing money, Gann, profiting from the bearish market, earned even more, because he was not against selling short. In fact, he advised his clients to sell short.

You might ask yourself, where did he start? There must have been an origin. We don't see life as a circumference without a center, or a mass concentrated in a unique center, like the Aleph of Jorge Luis Borges. Life is such a mysterious thing that to be understood, it needs to be organized and recapitulated, as Carlos Castaneda proposes in his books, based on the teachings of Don Juan. Where you find knowledge, you often find malice and mischief. Only wisdom remains immune, but it's not as easy to find.

People continue to be inspired by Gann and his techniques, applying his methods and principles; however, sometimes these same principles are not very clear to the traders applying them. This is because Gann, in his astonishingly expensive weekend private classes ($5,000 in the 1930s), taught the principles of his geometrical angles and the squares of nine and 12; however, he didn't explain anything about his more advanced techniques. Nevertheless, Gann's approach to the angles seems to be a linear one rather than a quantum one, as we propose later on.

We devote the next section to understanding how the principles that led to Gann's success were developed.

GANN'S THEORY OF STOCK MARKET CYCLES

The majority of people believe that markets and stocks follow random and irrational trends. Even great economists have often argued that financial markets are guided by erratic forces. They say that it's more appropriate to talk about a *random walk* since no one is able to foresee the trend of the markets, given that many exogenous events cause markets to be nervous and volatile.

Gann accepted these elements, but was convinced that markets move according to cyclical laws originating from mathematics and science. He thought that financial markets followed models and paradigms in which exogenous variables could be explained by metamodels with high heuristic and forecasting potential. Gann's analytical models had a rather ambitious goal: to precisely define the area of reversal of a trend and spot the tops and bottoms of single stocks and commodities. This was his favorite sport and he practiced it religiously, showing off his trading performance to those who followed closely.

He dedicated 10 long years to diligent research, going from library to library, from region to region and from state to state, until he was sure of his discovery. "The Vibration Law," which explains the movements of stock prices in a scientific way, would turn out to be the biggest accomplishment of his life.

Gann was one of the first to represent prices graphically in a chart with Cartesian axes, with time on the vertical axis and price on the horizontal one. This made it possible for him to not only detect graphical figures of support and resistance, but also to identify asynchronous cycles of diverse sizes. These cycles repeated themselves, allowing for reliable forecasts and, most important, profits.

In each library he visited, besides gathering data on the relevant commodity prices, he also read books on mathematics, physics, astronomy, and philosophy. Thanks to his photographic memory and his intellectual brilliance, he could memorize and synthesize large amounts of information, organizing it systematically and creating unconventional, analogical, and interdisciplinary links.

"The Vibration Law" is the synthesis of analyses of price cycles on infinite historical series over time, with the application of scientific principles and laws of physics to movements in stocks. For Gann, prices were like particles that move in space and time, and jump from one price level to another, like an electron.

It shouldn't come as a surprise that Gann loved Aristotle and that he saw new information, or rather, *rumors*, as secondary causes to price movements, whereas many traders held that *rumors* were the primary drivers behind movements in stock prices.

Take a look at the following interview with Gann by Arthur Angy, released to *The Morning Telegraph* on December 17, 1922. It contains some very interesting insights in understanding Gann's approach to the markets.

Wall Street Scientist Forecasted Top of Bull Market One-Year in Advance: His Indications Uncanny

W.D. Gann has scored another astounding hit in his 1922 stock forecast issued in December 1921. The forecast called for first top of the bull wave in April, second top in August, and the final top and culmination of the bull market October 8 to 15, and strange as it may seem, the average prices of twenty industrial stocks reached the highest point on October 14 and declined 10 points in 30 days after that date.

Mr. Gann predicted a big decline for the month of November. He said in the 1922 forecast, "November 10–14 panicky break." During this period stocks suffered a severe decline, many falling 10 points or more in four days, and on November 14 lowest average prices were made with 1,500,00 shares traded in on the New York Stock Exchange.

I found his 1921 forecast so remarkable that I secured a copy of his 1922 stock forecast in order to prove his claims for myself. And now, at the closing of the current year in 1922, it is but justice to say I am more than amazed by the result of Mr. Gann's remarkable predictions based on pure science and mathematical calculations.

The North Side News stands for a clean Wall Street and has rendered a great public service in helping to rid Wall Street of the bucket shop evil by publishing a series of articles in conjunction with the Magazine of Wall Street. *We believe in banding a fake, and we believe in giving credit where due.*

GANN IS NO TIPSTER

W. D. Gann is no "Wall Street tipster" sending out market letters and so-called-inside information—Mr. Gann's results are obtained by profound study of supply and demand, a mathematical chart of money, business, and commodities. He determines when certain cycles are due, and the order and the time when market movements will follow.

During the past thirty years many men have proclaimed discoveries and theories to "beat the Wall Street game," most of which resulted in loss to their followers. They could always tell by the chart just why the market did it after it happened. Mr. Gann's theory differs from the others in that he tells months in advance what stocks are going to do.

His forecast stated that some stocks would make high this year in April, some in August, and others in October—the month when he predicted the bull movement would culminate. Of a list of a hundred stocks; 30 made highest price in April and many declined, while others continued higher, 20 made high during August, and 50 made high of the year in October, from which the largest decline of the year has taken place.

His 1922 forecast indicated final tops on railroad stocks for August 14. The Dow Jones' averages on rails made high August 21 and reached the same average levels on September 11 and October 16, but did not exceed the high made in August, which was made seven days later than the exact date called for in the forecast.

HIS CHART IS A FACT

Stock Market accurate long-range forecasting, as W. D. Gann is doing, sounds almost unbelievable, and how he does it I do not know, but the writer does know that he does it. My attention was first called to his 1921 Market Forecast, in which he predicted stocks would be bottom in August, 1921, and advance to December, 1921. They did so. His chart or graph of the market one year in advance is a fact, and that the course of the stock market follows it astoundingly close is equally a fact.

Mr. W. D. Gann says the trouble with most chart makers is that they work with only one factor—space movements or charts that record one to two points up or down—whereas there are three or more factors to be considered, space, volume, and time. The most vital is time, and back of that is the cause of recurrence of high or low prices at certain intervals.

I asked Mr. Gann: "What is the cause behind the time factor?"

He smiled and said: "It has taken me 20 years of exhaustive study to learn the cause that produces effects according to time. That is my secret and too valuable to be spread broadcast. Besides, the public is not yet ready for it."

"Water seeks its level," continued Mr. Gann. "You can force it higher with a pump, but when you stop pumping it requires no force to cause it to return to its former level. Stocks and commodities are the same. They can be forced above their natural level of values to where lambs lose all fear, become charged with hope, and buy at the top. Then stocks are permitted to sink to a level where hope gives way to despair and the most rampant bull becomes a bear and sells out at a loss. My discover of the time-factor enables me to tell in advance when these extremes must, by the law of supply and demand, occur in stocks and commodities."

"We use the square of odd and even numbers to get not only the proof of market movements, but the cause."

Now, let's examine another famous interview released by *The Ticker and Investment Digest* in 1909. In the interview, Gann described his groundbreaking findings:

In going over the history of markets and the great mass of related statistics, it soon becomes apparent that certain laws govern the changes and variations in the value of stocks, and that there exists a periodic or cyclic law that is at the back of all these movements. Observation has shown that there are regular periods of intense activity on the Exchange followed by periods of inactivity.

Mr. Henry Hall in his recent book devoted much space to "Cycles of Prosperity and Depression," which he found recurring at regular intervals of time. The law which I have applied will not only give these long cycles or swings, but the daily and even hourly movements of stocks. By knowing the exact vibration of each individual stock I am able to determine at what point each will receive support and at what point the greatest resistance is to be met.

Those in close touch with the market have noticed the phenomena of ebb and flow, or rise and fall, in the value of stocks. At certain times a stock will become intensely active, large transactions being made in it; at other times this same stock will become practically stationary or inactive with a very small volume of sales. I have found that the law of vibration governs and controls these conditions. I have also found that certain phases of this law govern the rise in a stock and an entirely different rule operates on the decline.

While Union Pacific and other railroad stocks that made their high prices in August were declining, United States Steel Common was steadily advancing. The law of vibration was at work, sending a particular stock on the upward trend whilst others were trending downward.

I have found that in the stock itself exists its harmonic or inharmonious relationship to the driving power or force behind it. The secret of all its activity is therefore apparent. By my method I can determine the vibration of each stock and also, by taking certain time values into consideration, I can, in the majority of cases, tell exactly what the stock will do under given conditions.

The power to determine the trend of the market is due to my knowledge of the characteristics of each individual stock and a certain grouping of different stocks under their proper rates of vibration. Stocks are like electrons, atoms and molecules, which hold persistently to their own individuality in response to the fundamental law of vibration. Science teaches that "an original impulse of any kind finally resolves itself into a periodic or rhythmical motion;" also, just as the pendulum returns again in its swing, just as the moon returns in its orbit, just as the advancing year over brings the rose of spring, so do the properties of the elements periodically recur as the weight of the atoms rises.

From my extensive investigations, studies, and applied tests, I find that not only do the various stocks vibrate, but that the driving forces controlling the stocks are also in a state of vibration. These vibratory forces can only be known by the movements they generate on the stocks and their values in the market. Since all great swings or movements of the market are cyclic, they act in accordance with periodic law.

Science has laid down the principle that the properties of an element are a periodic function of its atomic weight. A famous scientist has stated that "we are brought to the conviction that diversity in phenomenal nature in its different kingdoms is most intimately associated with numerical relationship." The numbers are not intermixed accidentally but are subject to regular periodicity.

The changes and developments are seen to be in many cases as somewhat odd.

Thus, I affirm every class of phenomena, whether in nature or on the stock market, must be subject to the universal law of causation and harmony. Every effect must have an adequate cause.

If we wish to avert failure in speculation we must deal with causes. Everything in existence is based on exact proportion and perfect relationship. There is no chance in nature, because mathematical principles of the highest order lie at the foundation of all things. Faraday said, "There is nothing in the universe but mathematical points of force."

Vibration is fundamental: Nothing is exempt from this law. It is universal, therefore applicable to every class of phenomena on the globe.

Through the law of vibration every stock in the market moves in its own distinctive sphere of activities, as to intensity, volume, and direction; all the essential qualities of its evolution are characterized in its own rate of vibration. Stocks, like atoms, are really centers of energy; therefore, they are controlled mathematically. Stocks create their own field of action and power: power to attract and repel, which principle explains why certain stocks at times lead the market and "turn dead" at other times. Thus, to speculate scientifically it is absolutely necessary to follow natural law.

After years of patient study I have proven to my entire satisfaction, as well as demonstrated to others, that vibration explains every possible phase and condition of the market.

Myth or reality? The astonishing results achieved by Gann throughout his career and the numerous articles about him suggest that he really was a trading master who understood that stocks and commodities are ruled by the same principles at the base of quantum physics laws. To be successful in the financial market you need a nonlinear approach.

The way W. D. Gann presented his forecast is typical of a nonlinear approach because he speaks about the price in a way very similar to an electron's behavior, as we can see in the next quotation from the previous article:

He came to me when United States Steel was selling around 50, and said, "This steel will run up to 58 but it will not sell at 59. From there it should break 16 points." We sold it short around 58 with a stop at 59. The highest it went was 58. From there it declined to 41, 17 points.

When the price was 50 and the trend was up, Gann predicted that the target was 58, and at that level he would sell short the stock to cover it 16 points lower, at 42. In his calculation 58 was the resistance, corresponding to our concept of a higher QPL, and 42 was the support similar to our concept of a lower QPL, as we have already studied in Figure 5.2 in Chapter 5. Unfortunately Gann never explained how he calculated these support and resistance levels, only explaining simpler techniques like price retracements and double and triple tops and bottoms. His explanation of Gann Angles has nothing to do with price's quantum behavior

Even though he never publicly explained how to apply his ideas, he was very inspiring for me. I developed my Quantum Trading model based on the main concepts of quantum physics after reading his few, but meaningful, words about stocks, atoms, and electrons.

CHAPTER 7

Chaos Theory and Gann Angles

A butterfly, flapping its wings in New York, moves an infinitesimal mass of air, and this, through a series of chain reactions, leads to a hurricane in Miami causing serious damage. Obviously this is a paradoxical situation, but it is a form of reasoning that uses the latest metamodels derived from contemporary physics and, more precisely, it is the reasoning at the base of chaos theory. One aspect of chaos theory studies series and chains of phenomena that lack a cause-effect relationship. The aim of chaos theory is to develop models able to create functions that can mathematically describe and represent the statistical data derived from phenomena apparently heterogeneous and casual.

Ok, let's get back to the butterfly. Have you ever seen a butterfly flap its wings in New York and then cause a hurricane in Miami? If you really answer yes, then only you in all the world has been able to witness this!

And yet you have probably heard of avalanches falling from mountains into valleys and causing serious destruction. In the case of the avalanche, what do you think has happened? Actually, something similar to a butterfly's flapping wings causing a hurricane in Miami. Very often an avalanche begins with a bit of snow at the top of a mountain. As the snow slides down it picks up velocity and encompasses larger and larger quantities of snow until it becomes an avalanche of gigantic size. If you knew that the snowball was destined to transform itself into an avalanche, you would evacuate the targeted area in time.

Why can't you know this ahead of time? It's because, despite the fact these phenomena have a temporal continuity, you see only the final event, the avalanche. There is a temporal continuum that links together events with cause-and-effect structures, which are extremely precise and deterministic. Unfortunately, you do not know the beginning of the story; you don't know the first ring in the chain of events that ends up as an avalanche.

If we had a helicopter from which we could follow all of the phases from the first shift of a handful of snow to the avalanche, we would see everything from the beginning until the catastrophic final impact. We could film the entire flow of events to see the changes from moment to moment.

Instead, chaos theory operates with a series of data, apparently uncorrelated, and aims to provide mathematical models able to predict the final events.

The difference between the famous butterfly example and a snowball transforming into a fatal avalanche is that the air moved by the wings of the butterfly is apparently too small a phenomenon to be observed. Furthermore it is invisible and it cannot be filmed, while the event of an avalanche could, in theory, be filmed from beginning to end. Gann was a master *ante litteram* of chaos theory: He measured the effect of imperceptible movements, like the flapping of the butterfly's wings, which can have extraordinary influence on the financial markets. Gann Angles enable us to measure events that are spread out in time and space. They help us in determining how price increases mass and momentum, and in which direction.

Through the medium of Gann Angles one goes from a negative infinity to a positive infinity. When time and price come into equilibrium, even after decades, we can see in the markets a major reversal. Let's check it out and see how it works.

HOW GANN ANGLES ORIGINATED

The base of Gann Angles is the circular dodeca-partite algorithm (CDPA). Dodeca-partite comes from Latin and means divided into 12 parts. Even though Gann never spoke about CDPA, I consider it the real origin of the trading tools that every trader knows as Gann angles.

Below I explain the main features of Gann Angles. I don't claim to exhaust all issues and properties connected with Gann Angles. Rather, I introduce the most important concepts to study and use Gann Angles in a quantum way.

CDPA can be basically represented as a circle on which two forces move around in opposite directions. These two forces are price (P) and time (T) (see Figure 7.1).

The circle is the geometrical figure that conveys the concept of "wholeness," and price moves throughout time. When price and time are in balance with each other we can have a reversal. Gann noticed that a top or

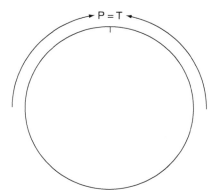

FIGURE 7.1 The Base of Gann Angles

a bottom can occur when price squares time. So we can come back to the theory of relativity, but instead of using Einstein's concept of space-time, we can tap into the price-time concept, which is more useful with respect to our trading goals.

Now, just to give you a rough idea of what we mean, if the price of a stock moves from 113 to 133 in 20 days, we are facing a time-price equilibrium and the two forces on CTDA balance each other out. Gann would say that the stock moved 20 dollars in 20 days and so it's ready for a reversal. In my opinion, to have a top or a bottom, the equilibrium between time and price is a condition necessary, but not sufficient. I discuss this subject more extensively further on.

We can define Gann Angles as a price-time vector because by drawing them from a significant top or bottom, we can have a clear understanding of how price develops throughout time.

When price meets time the market will experience a moment of stasis, and we'll see a trend reversal in the chart.

As we have already seen in Figure 7.1, two forces (P and T) rotate. If you want to graphically depict where these two forces meet and create equilibrium, in order to spot on your favorite chart where a reversal can occur, we need to proceed as follows.

We divide the circle into four quadrants, as we can see in Figure 7.2, and we redraw the upper-right quadrant with two orthogonal line segments. As a result, the quadrant becomes a square, as shown in Figure 7.3. This square is used to draw a Cartesian coordinate system, with time on the x-axis and price on the y-axis. Now we can draw a 45-degree line bisecting the square, as you can see in Figure 7.4.

Every point of the bisecting line identifies, by definition, a point where price and time are in equilibrium, because each point of the bisecting line

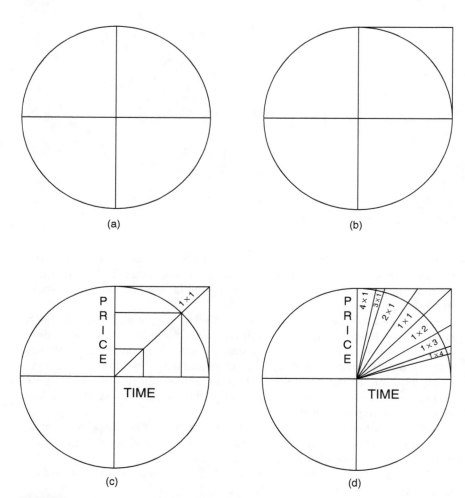

FIGURE 7.2 Price and Time in Equilibrium

corresponds to equal segments on both the x-axis and the y-axis. This concept belongs to geometry 101 and is very easy to understand.

Now let's draw on either side of the bisecting line, as shown in Figure 7.5, the other segment lines that proportionally divide the square. This fan-shape of lines was called by Gann "my angles," and therefore is known by traders as Gann Angles.

Gann called the bisecting, 45-degree angle a 1×1 angle. This means that in each point placed along the line, the price rises (or falls) one unit of price for one unit of time.

FIGURE 7.3 S&P 500 Big Bottom and Gann Angles, 1987

The angle immediately above the 1 × 1 angle is called the 2 × 1 angle, and it means that in each point placed along the line, price rises two units of price for one unit of time.

The next-higher angle is called the 3 × 1 angle, and at any point along this line, price rises three units of price for one unit of time.

The angle immediately above the 3 × 1 angle is called the 4 × 1 angle, and at any point along this line, price rises four units of price for one unit of time.

FIGURE 7.4 S&P 500 Index Major Bottom and Historical Top Shown by Gann Angles

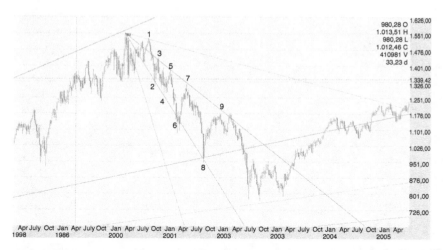

FIGURE 7.5 S&P 500 Index: Nine Consecutive Trading Signals Using Gann Angles

Gann Angles are also drawn below the initial bisecting line, in the same sequence as the preceding angles. The angle immediately below the 1×1 angle is called the 1×2 angle, and at any point drawn along this line, price rises one unit of price for any two units of time.

The next-lower angle is called 1×3 angle, and at any point drawn along this line, price rises one unit of price for any three units of time.

The angle immediately below the 1×3 angle is called the 1×4 angle, and at any point drawn along this line, price rises one unit of price for any four units of time.

As you probably have already noticed, every point of the 1×1 angle represents an equilibrium between price and time. It means that, when the price touches the 1×1 angle, we have an equilibrium, and therefore we could expect a potential reversal in the chart of the security. Actually, as we mentioned before, this is only a sufficient, but not a necessary, condition, and so to open a new long or short trade we need confirmation from our QPLs, the most important in the hierarchy of the trading tools that we use.

Also, every point along the lines of the other angles show an equilibrium between price and time, although it is not based on a 1:1 ratio anymore, but rather on different ratios such as 1:2, 1:3, 1:4, and so forth, or vice versa. What we have just said in the previous paragraph about a potential reversal that can occur on 1×1 angle is true also for all of the other angles. Always consider that we are speaking about a condition that is necessary, but not sufficient.

This means that a Gann Angle can be seen as a resistance or support line with very special characteristics. They are like invisible and immaterial walls that can be perforated by the price, and they become solid only when they're activated by QPLs. In this case they become very strong levels of support and resistance, as we will see in the next section.

In this example we show ascending angles. We usually draw ascending angles starting from a bottom. The origin of the Gann Angles will overlap with the bottom. In case we want to study the price behavior starting from a top, we have to do the opposite, overlapping the origin of the angle with the top. In this case we'll draw descending angles.

Gann also called angle 1×1 a 45-degree angle, and the 2×1 angle the $63^3/_4$-degree angle, and so on. Notice that these geometrical correspondents are true only when the chart of your stock or currencies is squared, that is, as if you are using graph paper, and the length of a unit of price is equal to the length of a unit of time. In this case the 1×1 angle is always the bisector line of a square.

Because we use software that helps us to display Gann Angles on charts that fit a computer screen, unfortunately our charts are almost never a square chart, like the chart that Gann used to draw by hand on graph paper. Furthermore, there is some software that displays 45-degree angles, but these may have a different appearance because they can be distorted by the scale you use on your computer screen. The 45-degree angle (1×1) displayed by Gann Angles software doesn't change if you change the scale of the screen.

Another important concept to study is the slope of the 1×1 angle, which is very similar to the concept of CS (conversion scale), as Chapter 5 demonstrates for the Price Space Operator (PSO).

If you have a stock ranging from 100 to 200 you can use 1 as the slope for angle 1×1. It means that each day the price, according to this angle, rises or falls 1 point, which can be equal to 1 dollar. But if you have a currency that ranges from 1.10 to 1.50, the slope of angle 1×1 must drastically change; otherwise it would be useless. In this case you will use as a slope for this angle another value, different from 1. Different authors and traders propose, in this case, several methods to calculate the 1×1 angle.

The only method to calculate the slope we use follows a "quantum" scale and stems from the same approach we used for calculating CS value. The number we use as a multiplier can be found in only one of the following sequences:

1, 2, 4, 8, 16, 32, and so forth—that is, equal to 2 to the n power $^1/_2$, $^1/_4$, $^1/_8$, $^1/_{16}$, $^1/_{32}$, and so forth—that is, equal to 1 divided by 2 to the n power.

As you can see here, we are also using a "quanta" factor to calculate the slope of angle 1×1.

Whoever is familiar with the Gann Angles system has already understood the importance of this subject. This is crucial for a correctly drawn Gann fan.

So what we propose is using only one of the above numbers to calculate the slope of the 1×1 angle. For instance, if you take the EUR-USD chart, I would use 0.003906, which is equal to $1/256$, one of the numbers of $1/2$ to the n power.

If you set the slope of 1×1 angle, the other angles of the fan are automatically set.

There are many special properties of angles 1×3 and 3×1, but this is not a book dedicated completely to Gann Angles and techniques, and it would not do to discuss in detail topics such as commutative properties, translative properties, and time signals.

Nonetheless, it is worthwhile to briefly state that if you rigorously use the 2 to n power quantum scale, according to the translative property, all of Gann Angles are interchangeable (except for angles 3×1 and 1×3, which follow another rhythm of growth). You do not have the problem of calculating which factor to multiply or divide to find the slope of angle 1×1. It can equally be 1, 2, 4, and so forth, because the angles, using the "quantum" scale, are proportional and interchangeable. The only thing to observe is that if the slope is too steep and the majority of the chart is not covered by the angle, you can use the next value in the 2 to n power scale to shift the angle to another position. So angle 1×2 becomes angle 1×4, angle 1×1 becomes angle 1×2, and so on.

Finally, if you take a fan of Gann Angles and you transport it to your favorite stock chart, you will obtain many price-time levels of supports and resistances. Just remember to always let the angles start from a top or a bottom.

Let's look at the following example, where we study the S&P Index chart with Gann Angles (Figure 7.3).

THE FALL OF THE STOCK EXCHANGE IN 2000

It is very important to draw the angle from a very prominent top or bottom from which the market has developed a significant dynamic momentum.

One example of this concept can be represented by the very important bottom in the October 1987 stock market, which was the terminal point of one of the most dramatic crashes after 1929. American stocks, followed by the rest of the world, declined around 20 percent in one day, and within a few weeks lost over 40 percent of overall value. If we take an S&P 500 Index weekly chart and we start drawing Gann Angles from the October 1987 bottom, we'll discover how they can very precisely show you after 13 years the historical top of 2000.

As we know, the rule states that an angle can become a resistance or a support point under certain specific conditions (but not always in these cases, as most traders think). Whenever an angle is broken during a bull campaign, it means that a resistance is broken, and the market will tend to reach the price level pointed out by the next highest angle. After touching it, if it's unable to quickly break it, thereby continuing its upward trend, the price will fall, using the angle immediately below it as a support.

Let's start drawing the angles on the weekly chart of October 1987 (Figure 7.4), using the coefficient 2, to calculate the slope of angle 1 × 1.

After various swings the price finds support in October 1990 on Angle 1 × 4, at point A, and from there starts to climb toward the higher angle, Angle 1 × 2 as indicated by letter C in our example. Then, in October 1996, the price finally leaves Angle 1 × 2 (C) behind and from that point begins the wild ride upward.

The bull campaign lasted many years, but what exactly is the target price? At what price could a top occur and the reversal begin?

If you don't know QPLs, Gann Angles can be very useful to answer these questions, if you use our quantum approach to calculate the slope of the 1 × 1 angle first. We'll try to spot the next historical top with Gann Angles and then, in Chapter 9, which introduces the concept of entelechy, we'll understand how QPLs give us the same results. Because both tools simultaneously give us the same answer, we have a very powerful cluster of reversal signals, This is what I call a condition sufficient and necessary to forecast a top or a bottom.

Coming back to our S&P 500 Index chart, we will find that the historical top formed on the angle 1 × 1 (indicated by letter D in Figure 7.4) at 1552 in the last week of March. The ascending angles originated from the bottom of October 1987, that is, one of the most important points from which you can draw Gann Angles, because the market dropped about 40 percent in a few weeks. As you can see from the chart, as soon as the price touched angle 1 × 1 on point D, in March 2000, it started to plummet. The historical top formed exactly on angle 1 × 1. Immediately after, the price started to fall. It continued to fall until it made a bottom, finding support on the angle 1 × 2 (indicated by letter C), which lies immediately below angle 1 × 1 on September 19, 2001, a few days after the September 11, 2001 attack on the World Trade Center.

Many have argued that the markets fell due to the tragic events and the threat of Osama Bin Laden. However, if we were to apply the mathematical model of Gann and read a chart of the price, we would predict a price decline to the angle immediately below, even before the collapse of the Twin Towers. In addition, we have to consider that the market had already started to fall more than a year before Bin Laden appeared on the scene, after touching the resistance of Gann angles at point D.

The market, having touched angle C (in this point two angles cross: angle C and the descending angle F from the top) on September 19, 2001, starts to strongly bounce back.

Then the price meets Angle E, a descending angle originating from the historical top of March 2000. As a result, the path toward the top is blocked, and the price falls until it touches the lower bottom.

Studying descending angles that originate from the historical top (see Figure 7.5), we see a series of nine, very precise, consecutive winning trading signals. You could sell short a future on S&P 500 at point 1. You cover the short and buy long at point 2. You close the long position and sell short at point 3. You cover the short at point 4 and buy long, you close the long position and sell short at point 5. You cover the short at point 6 and you buy long. At point 7 you close the long position and sell short again. At point 8, two angles cross each other, and this is another amazing entry point. So, you cover here the short position and you buy long. At point 9, finally, you cover the long position and sell short.

We buy and sell and each time we can make money opening long or short positions using futures or options. Gann angles provide entry signals for both long and short positions on S&P 500 futures or options at the price levels we have just discussed. If you had bought and sold in each of these nine points, you would surely make a handsome profit, but only if you used the advised quantum factor to calculate the slope of angle 1×1, applying the 2 to the n power scale we also use to calculate QPLs.

At this point we can say we have verified not only the extraordinary efficacy of Gann Angles theory, but also how Chaos Theory models work. In fact, the bottom in 1988 corresponds to the initial event from which a vibration stems, similar to the butterfly flapping its wings in New York. When the mass of air moved by its wings is magnified, according to the mathematical model identified by Gann Angles, as the price touches the last available angle, a historical top forms. Immediately after, the price collapses and this corresponds to the catastrophic event represented by the hurricane in Miami. Using Gann angles together with QPLs, you will be in position to successfully trade the markets and make a significant amount of money.

CHAPTER 8

Money Management Strategies

M any traders develop good strategies to trade the markets, but even if they are successful for awhile, in the end they lose what they have earned as well as their capital. According to various statistics reported by different authors, it seems that about 95 percent of traders lose everything between six months and one year after beginning, and only 5 percent survive in this wild jungle called the market, making consistent profits over a 10-year period.

The reasons for this premature departure are many: lacking discipline in controlling risk, using a too-high leverage, and not using stop-loss. In two words, lacking a money management method to rule our trades is a "fatal mistake." These, and other reasons we'll discuss further on, are what cause a trader to lose everything. Capital is the trader's life-blood, and if he doesn't preserve it carefully, he risks his entire investment.

I've seen many traders who were quite successful at the beginning, and then abruptly stopped trading forever because the market sucked out all of their blood, just like a thirsty vampire. Have you seen *Twilight*? Or Bram Stoker's *Dracula*? When you approach trading you're not approaching only the market, you are approaching the Prince of Darkness, the Lord of Vampires. You need to be on your toes. You need to be smart and remember that if you don't place a stop-loss, then the market could take all of your capital.

Before creating your trading action plan, it's very important to thoroughly understand how to take and control losses. When people asked W. D. Gann how to make profits in the markets, he usually answered that if a trader wanted to make gains, he would first have to understand why there are losses. He added that when you are able to control risk and minimize your losses, you can then start thinking about gaining in the markets.

WHY SOME TRADERS HAVE BIG LOSSES

The following are the four principle reasons that traders have losses in the market.

1. Lack of money management.
2. Lack of the use of stop-loss.
3. Use of average loss—the biggest mistake for a trader.
4. Making investment decisions by relying solely on *rumors*, which often end up cheating you, instead of focusing on the main causes that contribute to price fluctuation in the stock market.

Let's now take a more in-depth look at these four points. When people invest the majority or even their entire capital in only one operation they take on enormous risk. They fail to control their capital because they are governed by greed, wishing to realize huge profits in as little time as possible and ignoring the most elementary rules of prudence.

Many beginners in the Forex market are attracted to the apparently appealing proposal of online brokers who offer leverage up to 1:200. This means that having a balance in the account equal to $10,000, they can trade up to a countervalue of $2 million. If you use all of your maximum leverage for one trade, you can lose all of your capital with an adverse market movement of only 0.5 percent. In this way the stock market begins to look more like a bet at the casino or the horse track where you hope luck will be on your side.

Instead we need a money management system with strict rules, which, when followed correctly, allows a trader to rest well at night.

What is stop-loss? It's a selling order (or buying order if you have a short position) that allows you to exit the market automatically when it starts going in the opposite direction from the one forecasted. Stop-loss is like a life jacket, and you need one to survive in the rough waters of the investment sea. Now, we'll proceed to examine different issues related to stop-loss and money management.

The first step is to understand what your risk tolerance is for each trade. If you are conservative, you would risk only 1.5 percent for each trade. If you are more aggressive you can risk more, for instance from 2 to 3 percent. In the last case it means that before you lose all your capital, you would experience 33 consecutive losing trades. Instead, if you follow an effective trading system, you can use a 3 percent stop-loss to get a potential gain of at least double that amount.

It is very important to first calculate your maximum tolerance of risk and then, on this basis, decide what size of contract to use in the trade. Let's look at an example on the S&P 500.

Assuming that you want to trade mini–S&P 500 futures, where one full point is equal to $50, and your risk tolerance for each trade is 3 percent of your trading capital (which totals $10,000), you need to proceed as follows: calculate how many contracts you can trade and how large, in terms of points, your stop-loss will be.

Three percent of your trading capital is equal to $600. If you use a stop-loss of 12 points, then you will trade only one mini–S&P 500 future contract, because $50 \times 12 = 600$. A 12-point stop-loss could be good for opening a swing trade position if you have a target of 35–40 points. Most of the time, I use a stop-loss of 7 points if volatility is low, and a 10–12 points stop-loss if volatility is higher.

On EUR-USD, I usually use a stop-loss of 23 pips if my target is only 55–60 pips, and a stop-loss of 35 pips if my target is 100–200 pips or more. In this way, you can take advantage of following the entire swing from the beginning, if you open a position on a QPL.

For a shorter time-frame trading, such as on a 60-minutes mini–S&P 500 chart, you can have a target of only 10 points. In this case a stop-loss of 12 points would be too big, and you should use a stop-loss of only 5–6 points. In this case because you are still risking a maximum of 3 percent of your capital, you can double the future contract size and you can trade two futures for each trade instead of only one, because now your stop-loss is only half with respect to a 12-point stop loss. Your trade will be more predictable because you have a larger position always maintaining the same risk level.

If you want to risk only 1.5 percent for each mini–S&P 500 trade, you can use a six-point stop-loss on one future contract only, because your money management system will not allow a bigger one.

If you want to use a stop-loss of only three points to take six points of profit, it is fine. In this case, if you want you can trade two futures contracts in order to maintain a maximum loss of 1.5 percent with respect to your trading capital.

You can apply the same rule I used for mini–S&P 500 futures for your favorite stock. Assuming that your trading capital is still $10,000 and your maximum risk tolerance for each trade is still 1.5 percent, and you want to buy a stock that is trading at $100, because you have here found a support, then you can use a stop-loss of 100 basis points, and you can purchase

150 stocks using a 1.5 leverage. In case your stop-loss is caught by the market, you will lose only 1.5 percent of your initial capital.

It is very important that you always respect these guidelines and you always put a stop-loss accordingly. Remember that if you don't place a stop-loss, sooner or later the market will sweep you out.

THE DANGERS OF THE "AVERAGE LOSS" STRATEGY

Let's say it's the fall of 2007 and you meet with your broker and ask: "What can I buy?" The answer will most likely be: "Buy some Lehman Brothers mutual funds and other blue chips traded at around $100; these are seen as a good deal because they have a potentially big upside."

The stocks you purchased at $100 fall to $90. You start to worry and ask your consultant: "Why are the stocks you recommended falling?" And your consultant says, "Don't worry! Even children know that shares rise and fall in the stock exchange. There is no guarantee that a stock you purchased will immediately rise, but eventually it will soar."

So you return home, relaxed, thinking that you have an expert in charge of your investments. Then the share price falls to $80. Even more worried, you go back to the bank and ask about the negative trend. Again, the same answer: "Don't worry. The fundamentals are very good and it will regain its price. The stock is just going through a bad phase at the moment."

Nevertheless, you're still very worried, because you invested $50,000. At this point your broker pulls out his winning card and says "Buy again the same stocks, investing another $50,000 at the current price, $80. This way, when the stock exchange increases to the acquisition price, $100, not only will you have recovered everything you lost, but you will have earned at least $10 on every share. The average price will not be equal to 100, but rather to 90 (100 + 80 = 180 or 180/2 = 90)."

Actually, the use of average loss is a mortal trap. If you follow the average loss strategy and you happen to make gains as you use it, you'll start to think that this is the instrument that will always bring you gains when you buy stock. You'll start to think, "Even if the market falls, I will always gain with average loss when it rebounds." Consultants in the financial sector often advise following a strategy of average loss with the assumption that the stock exchange is destined always to rise again. Let's see!

The stock, however, instead of going up, continues to fall. The share price hits $60, and your worries transform into true anguish. Once again you show up at your consultant's office and say: "I'm desperate, what do I have to do? My $100,000 has now been reduced to $66,000—I have lost a good 34 percent of my capital!"

Your consultant, driven by his pride, tells you: "It's at this point where we really see if an investor has balls. If you sell right away, you will transform a potential loss into a real loss. Instead, since the share price is now down at a historic minimum level, and since it is a really good share, I want to give you some advice. Do as I do: I own an apartment and borrowed money from the bank, renegotiating my mortgage, and offering my apartment as collateral. Then I invested another $100,000 in the same stock. I made the average loss on such a historically low price level that I am sure, in three to six months, or at the most one year, I will recover everything I have lost, and in addition I will have a nice extra profit. Take my advice. This is the way people with experience achieve huge gains in the stock exchange in the long term."

If you follow your consultant's advice, you could be in a lot of trouble; this is what has happened to thousands of investors. Employees, traders, professionals, and blue-collar workers may even be persuaded to put down their properties as collateral. The result is that some stocks you purchase at a very high price don't necessarily come back to their initial price. If you tap into the deceptive promises of average loss and don't use stop-loss, you can lose a fortune.

NOISE, HOT RUMORS, AND RECOMMENDATIONS TO BUY

Gann noted many years ago that many investors, chronically hunting for news, religiously read journals, news bulletins, or asked the well-informed for tips on shares that could bring gains. Those experienced with the stock market by now have learned that all these rumors do not lead to gains, but rather are often the cause of great losses. "Hot" rumors, at one time released by a restricted, inner circle of users, have by now lost all of their importance, having already been exploited and accounted for in share prices. Relying on rumors generated by the media, after having already assisted in an important shift of the share price, the investor entering the market is often left empty-handed because the share price starts to fall, an unequivocal sign that the game is already over and done with.

And what can we say about analysts' recommendations? After familiarizing themselves with the balance sheets of companies, they make recommendations to buy. The credibility of these recommendations was rather strong up until the year 2007, but then something unexpected happened. A lot of people realized that news and recommendations to buy CDO, toxic derivatives, and overvalued stocks were totally devoid of a solid foundation. Even credit ratings, which had been considered a solid guide to true worth, began to appear suspect. In fact, the subprime and toxic derivatives bubble exploded, erasing trillions of dollars' worth of value. The rest is history.

We can easily conclude that it is much better to avoid the deceptive practices of average loss and to always use stop-loss to keep the gains you have earned, and to not jeopardize your trading capital.

CHAPTER 9

Entelechy

The Most Powerful Quantum Trading Concept

Aristotle upheld perfection and elegance as the pinnacles of achievement, and so he coined the term *entelechy* to indicate the condition in which the essence of being is fully realized or actualized; the physicist Fritjof Capra, in his book *The Web of Life*, uses this concept to indicate a vital force that motivates and guides an organism toward self-fulfillment.

Entelechy holds a very important role in Quantum Trading and shows us the unique points of P-Space, endowed with a higher degree of curvature where a reversal in a stock or in a currency is more likely to happen.

We believe that the markets show, most of the time, a state of disequilibrium. The reasons for this instability are found in the nature of the universe itself, and in the second law of thermodynamics and entropy. Equilibrium seems to be a very rare condition in the financial markets as well as in the universe, so other traders and scholars have proposed disequilibrium models in financial price behavior.

George Soros proposed the terms *dynamic disequilibrium, static disequilibrium,* and *near-equilibrium* to describe different conditions in financial markets. He distinguished near-equilibrium as a state where it is very easy to change the condition of the system, and far from equilibrium as a state where conditions are difficult to change. This concept belongs to his reflexivity theory, in which the biases of individuals enter into market transactions and they can potentially change the perception of fundamentals of the economy.

Soros helps us to take another step toward a *quantum* view of financial markets, yet I consider disequilibrium and equilibrium in financial markets from another point of view.

I approach price dynamics in terms of both *Space Curvature* and quantum leaps, caused by exogenous shocks that affect the price-photon, or price-electron. In P-Space, these two phenomena happen simultaneously, or they are just two different sections of the same "string." The space curvature is caused not by the bias of markets' participants affecting market fluctuations, but rather the curvature and the participation coemerge spontaneously. So the market, occasionally and for only a few instants, can be seen as highly efficient. Eugene Fama, from a completely different perspective, proposed an "efficient market hypothesis (EMH)," formulated in 1970, and argues that financial markets do not follow a predictable path, but instead can be described as taking a "random walk."

Our P-Space and Quantum Trading entelechy are an alternative to the notion of a random walk. They don't claim to predict anything, but only to show points of high probability for a reversal. Just as in quantum physics, we speak in terms of probability when we find the price, like the electron, at a certain location.

If you review the study published by the Federal Reserve of Atlanta on solar spot activity and stock returns (see discussion in Chapter 4), you can look at it using a different approach, depending on your mind set and preference for a linear or nonlinear weltanschauung. The study could suggest to you that this correlation would have to be ruled by a cause-effect relationship, or you could be inclined to believe that the phenomena are coemergent.

Coemergence is a very important concept for understanding and untangling the interdependence of complex systems in many fields, such as philosophy, epistemology, and psychology—including semantic and semiotic—as well as finance. It is a very powerful heuristic tool for understanding reality.

In Dzogchen philosophy, considered by some Tibetan schools as the pinnacle of Buddhist wisdom, *lhan cig skyes pa'i ma rig pa*—coemergent ignorance, intrinsic unknowing—represents a crucial watershed between ignorance, wisdom, and understanding the intrinsic nature of mind phenomena, or lack of knowledge of one's own nature. Ignorance is coemergent with our innate nature and remains present as the potential for confusion arises, preventing us from understanding the real nature of things.

In Jungian thought, coemergence is at the base of "synchronicity," which helps explain many so-called occult phenomena, such as premonitory dreams and uncanny coincidences. Coemergence is the bridge that leads us to perceive and explore the collective unconsciousness, as defined by Jung: the field containing all archetypes, the warp and weft of our known reality. Plato would probably agree with Jung, at least with the part related to Plato's own hyperuranium world of ideas and archetypes.

In P-Space, space curvature and quantum orbitals coemerge and exist simultaneously as different parts of a string. String theory is the latest revolutionary approach to physics proposed by Michio Kaku.

According to the quantum system, we can have the following types of entelechies:

- QPL Entelechy: a crossing of two QPLs or QPLSHs.
- Angular Entelechy: a crossing of two Gann angles.
- Mapping and Trapping Entelechy: a crossing of a Gann angle and a QPL or a QPLSH.

GENERAL PROPERTIES OF QUANTUM ENTELECHY

Entelechy in Quantum Trading is what we can consider a moment of supreme and rare equilibrium, when the P-Space curvature reaches its highest level and the probability of a security reversal is very high. If you wonder if all financial markets react equally with an entelechy, the answer is that some securities will reverse, but others will not and will instead accelerate their trend. In that case, what would be the use of knowing that the price will achieve an entelechy? It is very important because, instead of a reversal, the price continues to follow the previous trend. Entelechy represents a very promising entry level as the price will continue moving strongly toward the next orbital or QPL. If this happens, then usually the price accelerates its speed and the daily range grows exponentially.

Let's assume that the trend of a security originates from a bull campaign and is approaching a Quantum Entelechy. Even if you have sold short against the Quantum Entelechy and your stop-loss has been caught, you can reverse your position and buy long to follow the trend that will then, most likely, accelerate its speed.

QPL ENTELECHY

Let's start with the first type of entelechy. It forms in our P-Space and in security charts when two QPLs cross, creating a very strong support or resistance. Normally we can expect a major or intermediate reversal, depending on the structure of the entelechy, because the mass of two objects in the P-Space sum each other and create a higher curvature. The price is like a light particle that deviates from its initial path. It deviates because of light deflection, or better, price deflection activity.

Quantum Entelechy indicates the condition of something whose essence is fully realized, and where the curvature and the price climax are the protagonist: a mid-noon apex in the case of top, or a midnight inferior culmination in case of a bottom. This is always with the full realization of a process: the end of a trend and a rare case of higher equilibrium spreading to the financial markets.

Nothing influences anything else. Space curvature, light deflection, price deflection, and tops and bottoms in your favorite stock chart belong only to a nonlocal correlation, a higher level of category phenomena. They all belong to a cosmic entanglement, obeying not only what Bohm would call "a higher hierarchy or super structure," but also the universal laws of the "theory of everything."

I understand many scholars probably could not agree because they are not used to applying physics models to find support and resistance levels in a stock chart and because they don't trade the financial markets professionally. Don't forget that we are speaking about trading models inspired by physics and not systems of equations to discover other laws of physics. Also, many professional traders will find it difficult to accept trading models that are based on the theory of relativity and other concepts coming from quantum physics.

The watershed event of distinguishing between linear and nonlinear thought is still big and it is difficult to go nonchalantly from one to another.

In fact, a major reversal will be indicated by an entelechy occurring when two of the "P," "N," "S," or "J" QPLs cross and the price is there, tangent to the quantum price line. Also "M" QPLs can form powerful entelechies.

If the QPL is above the price, and price continues to rise from a bottom, then the quantum price lines forming the entelechy will act as a very strong resistance.

Many times entelechy acts as an invisible wall that materializes out of the blue to block a big bull campaign lasting months or years. Prices start to dramatically sink, as you can see in Figure 9.1, where you can find the triple entelechy formed by three QPLs significantly curving the P-Space and causing the S&P 500 index historical top.

"S" QPL, "J" QPL, and "M" QPL cross each other at the same time at point A, exactly where the S&P 500 index all-time top occurs. The price-deflection activity bows to Einstein's light-deflection phenomena and allows a Quantum Trader to place the trade of his or her life, closing all the long positions on the S&P 500 and opening short positions for a major change in the trend.

FIGURE 9.1 Triple Entelechy on S&P 500 Index and Historical Top, March 24, 2000

DOUBLE AND TRIPLE ENTELECHY POWERS

Usually an entelechy formed by one QPL and one QPLSH is considered strong enough to forecast a short-term reversal. An entelechy formed by two QPLs is considered more powerful and can cause a change in the trend in the medium term. But if you have the chance to find the price near a super-entelechy composed of three QPLs crossing each other at the same time, then you can expect a major reversal, such as a historical top or a major bottom.

In Figure 9.1 we are dealing with a super-entelechy. When the price approaches the triple-entelechy composed of "S," "J," and "M" QPLs at point A, it is as if an electron has lost all of its energy and it falls toward the inferior energy-level orbital, emitting a photon of light as it returns to the ground state from its excited state. In other words, one of the biggest and most powerful entanglements between QPLs, entelechies, and the S&P 500 Index price-reversal occurred on March 24, 2000, at 1538. In the same day the price reached first the "J" QPL at 1527.50, and after a little while it also touched "S" and "M" QPLs, passing at 1538.

The result of this is the all-time top in the S&P 500 index chart uncannily forecasted using Quantum Trading tools such as QPLs and entelechies. In this example we have used CS = 2, applying the simple rules we have already explained in Chapter 5.

I advise you to try to calculate on your own the value of these three QPLs forming this triple entelechy and drawing these lines on the S&P 500 weekly chart. You can simply put the value of each QPL at the beginning of every quarter (and of every month, for "M" QPL only), draw yourself the QPLs on the chart, and enjoy the excitement of discovering months in advance where the entelechy will take place as a strong target for the end of the bull campaign and the beginning of the bear side.

The opposite will happen with a QPL entelechy acting as a support. The price will unexpectedly and wildly bounce, while everyone believes that the drop will never end.

A major or intermediate reversal can be pointed out by the crossing of two of the following QPLs: "P," "N," "U," "S," "J," or "M." If the entelechy is formed by three QPLs, then we expect a very powerful and quick change in the trend.

As an example, the S&P 500 strongly rebounded on March 6, 2009, thanks to another super-entelechy of "J", "M," and "N" QPLs, as shown in Figure 9.2. At point A, the three QPLs cross simultaneously and one of the largest V-shaped bottoms ever seen occurs. Traders and investors were pessimistic after months of a severe bear market, and no one bought stocks, but rather sold out. After the price met the triple entelechy at point A, the stock market rallied in a great burst of enthusiasm, and the rest is history.

We bow once again to the price-deflection phenomena. I wish that Einstein were still around to see how I have used his ideas of light deflection. I would show him how an effective trading tool can be

FIGURE 9.2 S&P 500 Index Major Bottom, March 6, 2009

built based on the theory of relativity and the theory that space is curved because of the presence of mass!

ANGULAR ENTELECHY

Angular Entelechy refers to the crossing of two Gann angles. Like the entelechy discussed in previous sections, it is a powerful instrument, even though it shows only minor strength and frequency to create a reversal.

Sometimes when an angle coming down from a top crosses another rising from a bottom, even if the price is not tangent, it can provide a time signal for a minor temporary reversal of the trend.

The concept is very easy to understand. You just draw Gann angles on the chart and look for any angles that cross. Just remember to use a quantum approach to the scale and to set angle 1 × 1 by using discrete numbers belonging to the series (see Chapter 7).

Figure 9.3 shows an example of Angular Entelechy at point 8. (This S&P 500 Index chart is shown in Chapter 7, as well.) At point 8, two Gann angles cross. One is a descending angle and the other is an ascending angle, and the price is right there, ready for a reversal. We call this an Angular Entelechy, and as you can see it is a very powerful reversal pattern in Quantum Trading techniques. In fact, after touching the Angular Entelechy, the price soars: a Quantum Trader can make a lot of money if this tool is used correctly.

FIGURE 9.3   S&P 500 Index Showing Angular Entelechy at Point 8

MAPPING AND TRAPPING

Mapping and Trapping occurs when there is a crossing between a Gann angle and a QPL or a QPLSH. It is a curious way to apply different instruments stemming from diverse concepts, but it is another kind of entanglement that works well in our P-Space.

While you should already be familiar with QPLs and Gann angles crossing, this is the first time that we've mentioned Mapping and Trapping. I was inspired, in choosing this name, by martial arts, such as Wing-Chung, a Kung Fu school founded by a Buddhist nun from Shaolin temple and practiced in Hong Kong, Taiwan, and the south of China. This is because in Wing-Chung there is a special fighting technique, called chi-sao, to trap the adversary within close confinement. Similarly, with Mapping and Trapping, after spotting the weak points of our adversary, which is the market, we trap and defeat it.

The Gann angle could be considered like a theoretical vector stemming from a top or a bottom that, in the presence of a curvature point of strength, divides the P-Space area into two parts where the curvature itself occurs and shows the more sensible points—in terms of price and time—of the entire P-Space where a reversal is likely to occur. In these points the probability for a reversal is very high, but in case the price continues its trend and doesn't reverse, the target is given by the next QPL or QPLSH.

We call this entelechy "Trapping" because the QPL and Gann angle that cross can also be considered a powerful trap. The price is captured because its momentum is crushed by the action of the QPL and Gann angle which ally against it.

As we have seen in the examples provided in this chapter, entelechy can be a very useful Quantum Trading tool to spot reversal points from which a new trend can originate. If you study it and carry out your research on this topic correctly, you can get very important results in trading.

Entelechy is about P-Space curvature, the higher degree of curvature you can find in your favorite chart, where the price either reverses dramatically or continues its trend with a new strength. In the next chapter we will provide the reader with more examples of how this powerful tool can be combined with other very important Quantum Trading techniques.

Thanks to the combination of these instruments, we will be able to forecast accurately a change in the trend and make very profitable trades.

CHAPTER 10

Forecasting Tops and Bottoms Using QPLs and Subharmonics

In previous chapters we learn how to calculate QPLs and QPLSHs and explore the differences between various kinds of QPLs generated by different levels and degrees of curvature in our P-Space. We learn that entelechy conveys the higher magnitude and is one of the more powerful indicators of a trend reversal. It is demonstrated that some QPLs show major tops and bottoms and other QPLs show intermediate reversal points, where we can find very interesting highs and lows. Finally, other QPLs can show minor reversals. Now we will broaden our knowledge of how these different kinds of QPLs work.

Let's first recall the QPL hierarchy:

"P," "N," "U," "S," and "J" QPLs show the major tops and bottoms.
"M" QPLs show intermediate tops and bottoms.
"Su," "V," and "Me" QPLs show the minor reversals.

"P," "N," "U," "S," and "J" QPLs are used to spot the most important reversal points of a trend. A bull campaign is likely to end when the price touches one of these QPLs. On the contrary, a severe bear market could instead exhaust its trend after touching a very strong QPL. "M" QPL can often be found in entelechies forming on a top or a bottom if the other object composing the entelechy is "P," "N," "U," "S," or "J."

It's useful to recall that the price of a stock, or whatever other financial security, is like an electron, a particle that can jump from one orbital to another. The price-electron can behave in only two ways: it can jump away from the nucleus, which is similar to a price increase in the market, or it can jump toward the nucleus, which is similar to a price decrease in the market.

Because in our model, QPLs are like different electron orbits, when the price (considered as an electron) falls down from a top, we need to observe what it does immediately after touching a QPL, which is supposed to offer a support. If it starts rebounding, then it is like an electron that receives an additional energy quanta package and moves away from the nucleus, and therefore can be ready for a reversal.

We use the same approach as quantum physics regarding Schrö-dinger's cat, introduced in Chapter 1. The cat is in an indeterminate state, alive and dead at the same time until we open the box and observe. Like-wise, in Quantum Trading we wait until the price arrives at a QPL, and as it touches we observe the price action, that is we observe if the price reverses or not. In case it starts reversing (and we can use many time frame filters to confirm this trend reversal), we'll open a new trade in the direction of the supposed new trend. In the event that it breaks the QPL, we'll expect a strong continuation and acceleration of the old trend.

Quantum Trading is an approach that allows you to trade the market in two ways: trying to forecast bottoms and tops using a reversal approach, or using a trend-following approach where we can forecast the target price of a movement, that is the next level of the orbital where the price-electron should arrive as revealed by the next QPL. So, when the price has already begun to rise and touches a higher QPL or QPLSH, and instead of making a reversal it breaks this resistance level, we can buy long, expecting that the price will continue to rise until the next QPL or QPLSH, which is the target of this movement. Instead, when the price has already started to fall and breaks through a Quantum Price Line below it, we can follow the trend expecting that the price will continue to fall until the next lowest QPL or QPLSH.

FORECASTING A TOP

The method I will explain below can be applied to any security, such as single stocks, stocks indexes, commodities, currencies, bonds, or to the futures that have them as their underlying. Imagine that you want to try to forecast the next top or bottom in your favorite chart and you wonder how to deal with many different quantum price lines. The question is, how do we choose the right one among the different quantum price lines? It's very easy. We learn in Chapter 5 how to draw QPLs and QPLSHs, which you can also draw on your own favorite chart.

The order in which we place the lines is always the same. Just as listed above, we begin with the "P" QPL because it comes first in the sequence: "P," "N," "U," "S," and "J." Let's take a look at a chart of GBP-USD, one of the most important currency pairs.

If we place QPLs on our GBP-USD weekly chart, we discover something very interesting. We'll start drawing the "P" QPL and its subharmonics on the chart using a conversion scale (CS) equal to 0.0078125.

You remember from Chapter 5 that we can easily obtain the "P" QPLSHs by dividing into equal parts the range between one "P" QPL and the higher or lower "P" QPL. QPLSHs are subharmonics of the principal QPL.

It is possible to make the calculations by hand, but why spend the time if you can use software to calculate them? We can ask the software to calculate the position of "P" QPL for the beginning of every quarter from 2006 to 2008 and place the value on the GBP-USD weekly chart (see Figure 10.1), joining the different price values we obtain for the beginning of each quarter to create a line.

Let's assume we are on March 10, 2006. Let's draw the 45-degree "P" and "S" QPLSHs as well as "P," "J," "U," and "N" QPLs. We notice that GBP-USD is at 1.72 (point A, a triple entelechy on Figure 10.1 at the extreme right), just on 45-degree "P" QPLSH and "J" QPL. This is what we call a triple entelechy, and so we open a long position thinking that the price is waiting for a reversal and it will go up in the next few weeks and months. At point A, the 45-degree "S" QPLSH crosses the other two lines forming a

FIGURE 10.1 GBP-USD Weekly Chart (CS = 0.0078125)

triple entelechy, which is much stronger than a regular entelechy formed by only two lines.

At this point if we want to calculate the target of this movement, we'll take into consideration the next higher 45-degree "P" QPLSH, which in this case coincides with the "P" QPL (that is, the origin of all "P" subharmonics). A QPL is more important than a subharmonic (QPLSH), and so we can make our forecast saying that the target of an upward price movement will be "P" QPL passing at 2.08 on June 1, 2007, and at 2.09 on November 9, 2007.

As time goes by, the reversal is confirmed and we see a strong upward movement starting from point A and lasting around two and a half years. On November 2, 2007, we check the price and we realize that it is very close to two different QPLs.

Taking a look at Figure 10.2, you will discover that the rally started at point A and ended at point B (point A in Figure 10.2 is the same point A of Figure 10.1, even if we have taken out an "M" QPL, to avoid confounding the reader with too many lines). In fact GBP-USD historical top was made exactly at point B in Figure 10.2 on "P" QPL on November, 9, 2007, and on the same day the price also crossed the "J" QPL, reversing its trend on a very strong entelechy between "J" and "P" QPLs at point B, which is one of the most powerful reversal patterns in the Quantum Trading techniques arsenal. However, the downward movement began on the top formed by "P" QPL and ended on the "P" 45-degree QPLSH at point C. This is the main issue in our Quantum Price model in which we see the price as an electron jumping from a higher orbital level to an inferior one. The entelechy is also

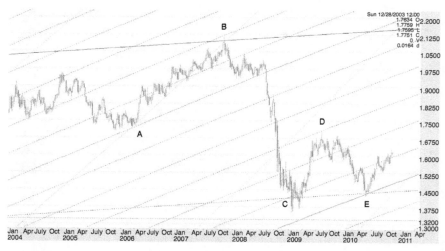

FIGURE 10.2   GBP-USD Weekly Chart Showing "P," "S," and "J" QPLSHs and QPLs

very important, but we want to be sure that the Quantum Price model is correctly understood.

In this case we are studying a currency chart, but please consider before you place QPLs on a stock's chart that you should make sure that you are not using a dividend adjusted chart. Instead, you should use an unadjusted one. The reason for using an unadjusted chart is because QPLs work with real prices that were actually exchanged in the market and not virtual prices obtained by modifying the real data, as happens for the adjusted charts. Please notice that the most popular free data providers usually offer adjusted charts and data.

The price of GBP-USD stock can be seen as both a light-photon or an electron according to the Quantum Trading model we introduce in Chapter 1. We begin by studying the price behavior as if it were a photon or a particle of light. In Chapter 1, introducing the parallelism between Einstein's theory of relativity and our Quantum Trading model, we notice that the light deflection phenomenon is crucial. The P-Space is curved by the mass of certain objects, and the price deflects from its path and changes direction, going from decent to ascent and vice versa (see Figure 1.5 in Chapter 1).

QPL Tolerance Rule

A QPL curves the P-Space not only at the precise point where the QPL is located on a certain day, but also in the surrounding area identified in the range from −1 percent to +1 percent, if you consider a daily chart.

If the price fails to touch a higher QPL and makes a trend reversal before touching it, it means that the price is in a weak position and the reversal can be severe. The opposite happens when the price fails to touch a lower QPL and a rally begins.

Sometimes the price will reverse after breaking a QPL of less than 1 percent, and this corresponds to the lost motion, which is the motion of a mechanism during which no useful work is performed.

Considering an intraday chart, the range of tolerance has to be reduced, according to volatility.

So GBP-USD's price can be seen as a particle of light meeting the "P" QPL and the "J" QPL at the same time. The curvature of the P-Space generated the reversal of the trend, and the price began to fall. That top price (2.11) hasn't been reached again to this day by this currency pair.

Coming to consider GBP-USD's price as an electron, it is easy to understand how the "P" QPL at 2.09 and "J" QPL at 2.09 (top at point B) is the target to which the price was attracted. In fact, the price arrives just at 2.11

after a long run-up. This QPL price corresponds to the next orbital level of an electron in the quantum physics model.

The next question could be, "Is there a deterministic relationship between the fact that QPLs are located at certain prices and the top or bottom share price has to reach it before reversal?" The answer is very similar to the one provided by quantum physics about Max Planck's constant (h): "We can only speak in terms of high probability that the price-electron could be found exactly there." Furthermore, we can observe that the relationship between prices, P-Space, and QPLs is based on the concept of entanglement or nonlocality.

In the same way, the probability that "P" and "J" quantum price lines would be reached by the GBP-USD stock price was quite high, but it wasn't guaranteed because we were dealing with a nonlinear model instead of a linear model.

If the price had broken through the two QPLs we would have expected the price to continue to rise until the next highest QPL located at 2.45 on April 11, 2008. Because "P" is a very slow object, within the next few months this quantum price line would also be very close to this value.

Instead, because the price after touching point "B," "P," and "J" QPL started to reverse its path, we considered that it could have touched a top. The next step to understand if it was really a top or, at least, a high, is to check the price-bar dynamic. If today's daily bar low is lower than yesterday's low, and tomorrow's daily bar low is lower than today's low, it could be a probable sign that the trend is changing. The more bars you have following this pattern, the larger the probability of a trend reversal. Following our model the next target would be the next lowest QPLSH, which passes at 1.75. In fact, this level was touched in the first week of September, 2008, and because this support line was broken, the price continued to fall to the next level at 1.41 in the last week of January 2009.

Another very interesting entry point to open a long position on GBP against USD was on May 19, 2010, at point E, when the price touched a triple entelechy between 90-degree "P" QPLSH, "S" QPL and "M" QPL (see Figure 10.2). After touching this very strong support the trend reversed and the rally is still up after six months today, in November 2010. We have not drawn "M" QPL on the chart, so as not to confuse the reader with too many lines, but if you try to calculate it on your own, you will find it right at point E together with the other two lines.

If you also display 45-degree "S" QPLSH and 45-degree "U" QPLSH in our GBP-USD chart, you will discover that they cross at point B in Figure 10.3, finding in this entelechy composed by two Quantum Lines a very strong resistance from which the price plunges. The bottom was also made at point A of Figure 10.3, where the lower "U" subharmonic offers a very strong support. I have proposed another set of subharmonics in a

FIGURE 10.3 GBP-USD Weekly Chart (CS = 0.0078125)

separate chart so as not to confuse the beginner with too many lines drawn on the same chart and to clarify that different sets of QPLs and subharmonics can work at the same time, giving a lot of information about the most important reversal points in a trend.

How to Choose the Right QPL to Forecast the Next Price Target

As a good rule of thumb, if a certain QPL or QPLSH showed itself to be a good support or resistance in the past, we can assume that another subharmonic of the same QPL will be a good support and resistance to forecast the next top or bottom. This happens very often.

The reader could ask himself why we have chosen to use a conversion scale of 0.0078125 and not a different one. Generally speaking, if the price of a stock or a financial security ranged from 0 to 100 in the last decade, we have to use a CS < 1. The possible values that our CS could be are only those obtained by using a "quanta" approach dividing 1 by the 2^n power.

The conversion scale (CS) for GBP-USD can only be one of the following values:

$$1/2 = 0.50$$
$$1/4 = 0.25$$
$$1/8 = 0.125$$

$$\frac{1}{16} = 0.0625$$
$$\frac{1}{32} = 0.03125$$
$$\frac{1}{64} = 0.015625$$
$$\frac{1}{128} = 0.0078125$$
$$\frac{1}{256} = 0.00390625$$
and so forth.

We choose $\frac{1}{128}$ because this CS offers many "price-electron orbital levels," or QPLs, that can be used in our forecasting to target the next movement of the price.

In particular, we choose this CS value to forecast the next top because the previous bottom was made on a significant combination of two QPLs. Let's assume we're in June 2006 and we want to forecast the next top. We look back at the price of a few months ago and realize that a bottom was made exactly on two QPLs crossing the price at the same time, the entelechy we have already seen. For this reason we decide that this CS value is able to explain the past price dynamics, and for the same reason we consider it as valid to make a forecast for the future price behavior. This is why we have chosen $\frac{1}{128}$ CS.

But if we had used a different CS, such as $\frac{1}{256}$, or 0.00390625, we would have found another very interesting set of QPLs. In fact, if we change the CS and obtain a different set of QPLs, it often happens that the new set of QPLs fits equally as well to forecast tops and bottoms. In our example, it is easy to see in the next figure (Figure 10.4), using 0.00390625 CS, the "P"

FIGURE 10.4   GBP-USD Weekly Chart, with a Change in the CS (CS = 0.00390625)

and "J" QPLSHs are on the top at point A, forming an entelechy, and "U" QPLSH is on the bottom at point B together with "J" QPLSH.

In fact, you can see how, at point B, "U" QPLSH and "J" QPLSH create an entelechy and offer a strong support to the price. From this bottom the price rebounds strongly. At point C, "P" and "S" QPLSHs cross, forming another entelechy offering another strong support for the next higher bottom. It is amazing to discover how the price of a security jumps up and down taking quantum leaps that can be understood through QPLs and the models I propose in this book.

Certainly, we have other means to strengthen our conviction that after touching the QPL, the price reverses its trend. We can use several time frames at the same time and so, coming back to our GBP-USD daily chart, we can follow the price dynamic also in the 60-minute bar chart. Another interesting point to notice is that, when following the Gann Angle theory, if we change the time frame, the structure of the angles changes accordingly. Comparing the Gann Angles starting from the same bottom in a 60-minute chart to the ones starting from the same bottom on a daily chart will lead us to conclude that the angle passes through different prices offering different support and resistance levels. Instead, the QPLs are the same for whatever time frame, assuming we are using the same CS.

FORECASTING A BOTTOM

The previous discussion has shown how to forecast a top, and it will now be easy to understand how to forecast a bottom using the same approach, but in a "specular" way.

Let's take a look at the EUR-USD chart.

In Chapter 3, we introduce only *how* QPLs work, without any explanation of the conversion scale (CS) or of the differences between QPLs and QPLSHs. Remember that every QPLSH is also a QPL, and in Chapter 3 we call, for brevity's sake, the subharmonic of a quantum price line a "QPL," instead of a "QPLSH." Now we must differentiate between the two types of quantum price lines.

In Figure 10.5, we place the 45-degree "N" QPLSH using CS = 0.001953125. At points A and B we see that the historical top and a major bottom are indicated by the contact between the price and the "N" subharmonics. Please notice how precise the reaction of the price is as it touches the "N" QPLSH at points A1, A2, and B.

In Figure 10.6, we draw both "P" QPL and "N" QPLSH at the same time. By using a conversion scale of 0.001953125 and drawing the QPL and QPLSH on the chart, it's easy to see that "P" QPL is located at 1.2290 in

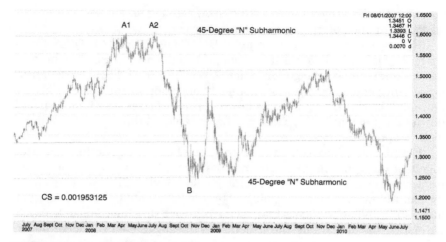

FIGURE 10.5 EUR-USD Daily Chart (CS = 0.001953125)

January 2008, and 1.2315 in October 2008. We use this conversion scale very often for EUR-USD studies. Notice that different securities respond better using a certain conversion scale instead of other ones, but you can simultaneously use more than one conversion scale if you want to find further confirmation, even though you are not forced to.

The euro finds a strong resistance at point D at 1.2461 where the "N" subharmonic passes. The upward movement ended at 1.5143 where

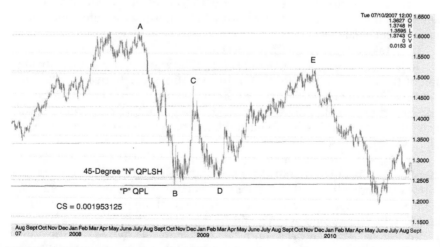

FIGURE 10.6 EUR-USD Daily Chart Showing "P" QPLs and 45-Degree "N" QPLSHs (CS = 0.001953125)

another "N" subharmonic passes offering a very strong resistance level at point E.

The "P" QPL remains more or less in the same place for a long period of time, and when the EUR-USD price, seen as a light-photon or an electron, touches this Quantum Price Line, it meets a very strong support. In fact, the price rebounds from point B at 1.2333 to point C at 1.47 in less than two months.

If you want to be more precise, you can display on the same chart 22.5-degree "N" subharmonics instead of the 45-degree "N" QPLSHs we have displayed before. In Figure 10.7, you can appreciate the accuracy of 22.5-degree subharmonics, which provide us with the resistance level at point C on December 18, 2008, from which the price plunged.

Coming back to point B, a beginner in Quantum Trading, after seeing the big rally occur after the price touches on the same day both "P" QPL and "N" QPLSH (Figure 10.6) could exclaim, "Wow! What a strong bounce! What happened?" Well, besides the consideration of fundamental data and the expectation of the market players for the economy's condition, in terms of Quantum Trading the price met one of the most important curvature points given by an entelechy composed by an "N" subharmonic and the "P" QPL that we consider as one of the most important indicators for a medium-term reversal in our hierarchy.

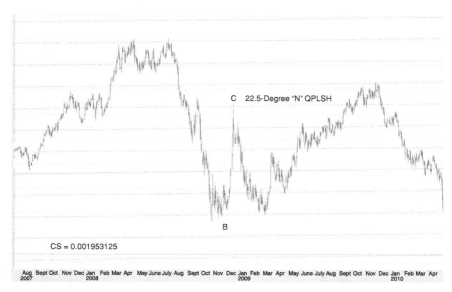

FIGURE 10.7 EUR-USD Daily Chart Showing 22.5-Degree "N" QPLSHs (CS = 0.001953125)

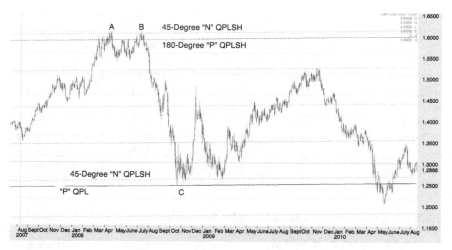

FIGURE 10.8 EUR-USD Daily Chart Showing 180-Degree "P" QPLSHs

If we consider the price as a particle of light, we can just say that it experiences the price-light deflection phenomenon that is at the base of our Quantum Trading theory. As we have already explained many times, if instead we consider it as an electron, when the price touches the "P" QPL and the 45 "N" QPLSH, it's as if it has received an external shock from which it receives more energy, and so it can jump to a higher level.

Now, let's take a look at the "N" 45-degree subharmonic in Figure 10.8 and we discover that on the previous double top, at 1.60, the "N" QPLSH passes just by there, and that also the 180-degree "P" QPLSH is close by, passing at 1.5819.

It's amazing to observe that the double top at points A and B, a plain vanilla reversal pattern very popular among the followers of technical analysis, in this case was not only confirmed by the 45-degree "N" QPLSH and 180-degree "P" subharmonic, but also anticipated before the second top occurred at the same level in point B by an entelechy. Even though our definition of an entelechy is the crossing of two QPLs, or QPLSHs, and in this case they do not cross, we still consider it an entelechy because the two lines are very close to each other.

So, QPLs and QPLSHs are the real indicators that allow us to forecast the all-time top that has never been broken so far. When a double top or bottom occurs on one, or more, very strong QPLSH(s) or QPL(s), it is the sign that the next movement will be steady and long lasting. This is what happens in this case because the EUR-USD top is done on the "N" QPLSH and "P" QPLSH. For this reason, we forecast that an important reversal will likely occur.

The 180-degree subharmonic is considered the strongest level after the QPL, and it is the next most important level to check to forecast tops and bottoms after a QPL, which is the origin of all other subharmonics. Chapter 5 describes how subharmonics are obtained as follows.

We start by taking the figure of the circle. If we take the circle as a unit, this corresponds to QPLs and their main harmonics.

If we divide the circle, which is composed of 360 degrees, by two, we obtain 180 degrees. This corresponds both to 180 degrees, or a 50 percent division of a range. This level is one of the most important to observe for a reversal, and you can also apply it to a simple static price retracement in a certain range stemming from a top and a bottom.

If we divide the circle by four, we obtain 90 degrees, which is equal to a 25 percent division of the range. If we divide it by eight, we obtain 45 degrees, which is equal to a 12.5 percent division of the range. Yet, the other subharmonics besides the 180-degree ones are also important, above all if they are very near to other subharmonics originated by a different P-Space object.

Coming back to Figure 10.8, the double top was made on both an "N" 45-degree QPLSH and a "P" 180-degree QPLSH. For this reason, we assume that when the price meets at point C, the level where the "P" QPL, a 180-degree QPL, is located, the price could be ready for a reversal. In fact, it will bounce on "P" QPL at point C, and this bottom can be seen as a major reversal point with a very high degree of probability. Furthermore, at point C we also find the 45-degree "N" QPLSH in addition to the "P" QPL. This is another important signal for a reversal: when a QPL passes close to a QPLSH. When the price of a security is very near to a QPL plus a QPLSH, we call this pattern Entelechy, one of the most important tools in our Quantum Trading model.

QPLSH RULE FOR 180-DEGREE ANGLES AND 90-DEGREE ANGLES

QPLSH Rule for 180 Degrees and 90 Degrees

As a simple rule, if a certain QPL has shown itself to be a good support or resistance in the past, we can assume that the 180-degree subharmonic of the same QPL will be a good support and resistance to forecast the next top or bottom.

If a certain 180-degree QPLSH has shown itself to be a good support or resistance in the past, we can assume that the same "object" QPL (which has originated the previous 180-degree QPLSH) will be a good support and resistance to forecast the next top or bottom. The same is true for a 90-degree subharmonic.

Of course we won't bet the entire farm, and we'll wait for the price bar dynamic to confirm our signal.

In fact, if you check on 60-minute charts and find a higher high on each bar and the QPL level is not broken anymore, it is an excellent confirmation that the signal is strong and valid.

The intraday models can be developed in the same way.

In Figure 10.9 you can see at point A how "P" QPL arrested the strong NASDAQ future uptrend on March 24, 2000. In this example we use a CS = 8. The price is tangent to the "P" quantum price line, and when it touches the line it suddenly reverses. The natural target is point B, the inferior QPL. Please notice that at point B, we don't find a QPLSH, that is, the origin of those subharmonics. The price then breaks the QPL at point B, and on a technical rebound the same QPL that was previously a potential support, because it was broken, now becomes the resistance. In fact, the price after retesting falls down, continuing its downward trend.

In Figure 10.10 we are using CS = 16 instead of CS = 8 as in Figure 10.9. We find on the same top at point A "N" QPL instead of "P" QPL as before.

FIGURE 10.9   NASDAQ Futures Daily Chart Showing "P" QPLs (CS = 8)

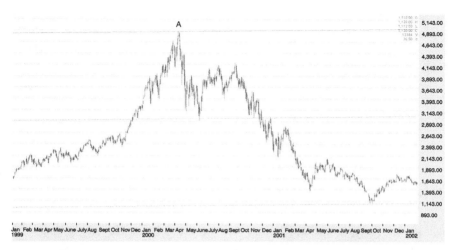

FIGURE 10.10 NASDAQ Futures Daily Chart Showing "P" QPLs (CS = 16)

This is to show that at a very important top, even though you change the CS scale you often find another QPL generated by the slowest object in our QPL hierarchy.

In Figure 10.11, we have come back to CS = 8, and we're still considering the NASDAQ future daily chart with the same historical top. The difference with respect to the previous figures is that this time we have put "P" 90-degree QPLSHs besides the usual QPL. As you can see at point B,

FIGURE 10.11 NASDAQ Futures Daily Chart Showing "P" QPLs and 90-Degree QPLSHs (CS = 8)

FIGURE 10.12 NASDAQ Futures Daily Chart Showing "P" QPLs and 72-Degree QPLSHs (CS = 8)

the trend finds a strong resistance after having tried to pull back, trying to reach the top again. Unfortunately the 90-degree QPLSH passes lower than the point A price level and the NASDAQ reverses, continuing to plunge inexorably.

Finally, in Figure 10.12, we find, still on the NASDAQ future daily chart, with the same initial CS = 8, but this time we have drawn the "P" 72-degree QPLSH. It is critical to notice at point D a 72-degree "P" QPLSH that arrested the severe NASDAQ plunge, which began at point C where a very strong "P" QPL stopped the uptrend, and offered a very strong support on the bottom that occurred on November 21, 2008, at 1016 (point D).

Behavior of the Price Near Strong QPLs

If another QPL generated by a heavy P-Space object is in the proximity of the previous one, we should wait for the price to reach the next QPL before opening a long position, because the price could reverse on the second QPL.

This happens very often, and we regard it as a rule of thumb.

The NIMEX crude-oil future chart is very interesting because it gives us the chance to show a Quantum Trading pattern happening quite often, so it is worthwhile to study it carefully. Looking at Figure 10.13, you can see that after a long run-up the price finally meets, at point A, the "P" QPL. It is

FIGURE 10.13   NIMEX Crude-Oil Future (CS = 0.5)

a very strong resistance level and the trend can reverse. As we mentioned before, when the price happens to be right on a QPL price level, that in this case can be a resistance, we just wait to see what will happen, exactly as Schrödinger would do with his quantum cat. We don't have a cat, but rather the price, and even if the price doesn't meow, it behaves more or less like our quantum cat. Therefore, in this instance we notice that the price breaks the "P" QPL and that it doesn't reverse.

At this point, according to our trading model, it could be profitable to open a long trade, if we were not already long before. But before opening a new trade, it's very important to check if there is some other QPL passing nearby on that same day. "P" QPL was a very strong resistance, but nonetheless, it was perforated. So now we have to verify if the rule mentioned in the previous box is proven true. We said before that if another QPL generated by a heavy P-Space object is very near to the previous one, we should wait for the price to reach the next QPL before opening a long position, because the price could reverse on the second QPL.

This happens very often. In fact, if you examine Figure 10.13, it is easy to see that the price, after breaking "P" QPL at 134.72, reached "J" QPL a few days later, passing at 143.77. The crude-oil price reached another very important resistance level on that day and "Su" QPL crossed "J" QPL. So we have an entelechy of the first type, generated by the crossing of two QPLs, as shown in Figure 10.14. According to our Quantum Trading model, it's the sign of a potential reversal of the trend and, like our favorite physicist, Schrödinger, we'll observe one more time the behavior of the price after it touches the entelechy at point A. The market tops and after a long, strong run-up, the trend reverses and a new bear trend begins.

FIGURE 10.14   NIMEX Crude-Oil Future and QPLs Entelechy

If you take a look at the entelechies, or the QPLs, that are very near to the present price where another QPL could be found, you can dramatically improve your ability to forecast price behavior according to the Quantum Trading model. Crude oil is a very volatile market, but if you can identify very good entry point levels, it is easy to make a lot of money because the trend of this commodity tends to be very precise.

Of course, you can decide to open only one third of your future position size based on the entelechy price, waiting to increase your position until after you receive confirmation that the trend is really changing from price-bar dynamic, as we have explained before.

At point A, the trend reverses and a bear market begins. Now we have to try to forecast the next price targets.

Applying the method we have used so far, that is very easy. Because the top was made on a "J" QPL, in order to spot the next support level where a change of the trend is likely to occur, we have to calculate the same object as a subharmonic. So we'll draw on the chart of future crude oil the "J" 120-degree QPLSH.

At point B in Figure 10.15, you can find the first 120-degree "J" QPLSH support, which is perforated. This is not a problem, because it is an indication that we have to continue to sell short crude oil, increasing the position because the trend is likely to reach the next "J" 120-degree QPLSH. The price continues to drop and crude oil extends its bear trend finally reaching the "J" 120 QPLSH. This passes at 33.05 dollars on February 12, 2009, and the price reaches it on this date (Point C in Figure 10.15). Immediately after, it reverses its trend, giving us the major bottom from the past

FIGURE 10.15 Nimex Crude-Oil Future and 120-Degree "J" QPLSHs

five years. From point C, a strong bull trend begins that is still going at the moment I am writing this chapter, with the current price at 86 dollars per barrel.

Let's take a look now at the CMX silver future chart (Figure 10.16). We can apply our Quantum Trading rules to whatever commodity chart is liquid enough to be decently traded without too much slippage. This chart covers a long period ranging from 2002 until 2010.

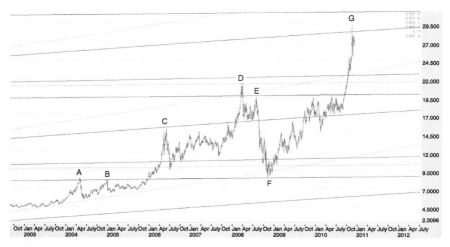

FIGURE 10.16 CMX Silver Future Daily Chart (CS = 0.03125)

For the first time we have drawn on the chart all of the slow-movement object QPLs, the most important ones in the QPL hierarchy. For the absolute beginner of Quantum Trading, it could be a little confusing to have all of the QPLs on the chart at the same time. So far we have shown QPLs one by one, but at this point the reader is surely acquainted with our method and so is ready to view on the chart all of our QPLs. In this case at point A, we find a very strong resistance generated by "P" QPL. At point B, we have a double top generated by the same "P" QPL. Then the price on the third attempt breaks this QPL and after a while also breaks "N" and "U" QPLs. Finally a strong resistance is found at point C, generated by "S" QPL. After breaking two QPLs the price of silver finally tops on "N" QPL at point D. At point E, "P" QPL is there and the price finds another strong resistance before continuing to drop toward the next "P" QPL, which offers a very strong support at point F. After touching the "P" QPL, the price reverses and a bull campaign begins, still continuing today.

The European Index DJ ESTOXX50 (Figure 10.17) can be traded profitably using Quantum Trading techniques. This index is composed of the average prices of 50 of the most important stocks in Europe. It is one of the largest volume-stock indexes in the world, together with the S&P 500 and Dow Jones, and it is largely traded by institutional and private traders all over the world.

The big rally that started on November 1999 met a tough resistance in March 2000. Of course many stock indexes topped that year in the same month, but as far as we're concerned, the bull trend was arrested by "N" QPL passing just at point A at the highest price that this market has ever

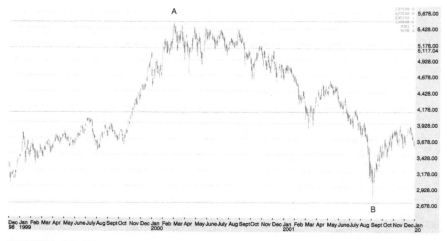

FIGURE 10.17 DJ ESTOXX50 and "N" QPLs

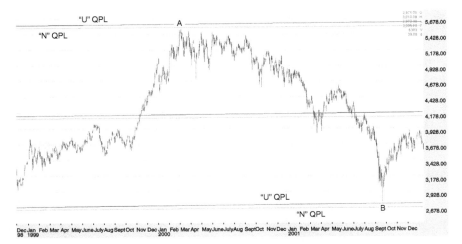

FIGURE 10.18 DJ ESTOXX50 "N" and "U" QPLs

reached. After touching "N" QPL, the bull trend ended its very strong run-up, and after some months it began to decline. The trend changed and a severe two-year bear campaign followed. At point B, the index found a very strong support in September 2001, very near the "N" QPL.

In Figure 10.18 you can find not only "N" QPL, but also "U" QPL displayed at the same time. In this case, the September 21, 2001, support is even more precise because the "U" QPL passes even more closely to the bottom, as you can see at point B.

CHAPTER 11

Time Algorithms in Quantum Trading

I n the preceding chapters we have learned to appreciate the heuristic potential of QPLs as a powerful tool able to give us important information about the targets of a trend in many kinds of securities. You can apply QPLs to any market that is listed, regularly traded, and has enough volume to be considered a "liquid market."

QPLs are mainly about price, the price where we can expect a major or intermediate reversal, meaning a top or a bottom, or if breaking a certain QPL, the next target price for a trend continuing to soar or plunge.

Now we will introduce a new concept in Quantum Trading regarding the use of P-Space to understand the "time" when a reversal is more likely to occur, regardless of the price. This "time" is calculated by using Quantum Trading Algorithms (TAs).

In this book we'll study only some aspects of absolute sinodical distances, and we'll not speak about the relative distance between two P-Space objects and other special time algorithms features, because the subject of TAs would require several books to be treated in a detailed way.

QUANTUM TIME ALGORITHMS (TAs)

The special angular distance between two objects of P-Space creates a singularity in time-space where the price is moving. This time-singularity can increase the degree of curvature of P-Space and generate a reversal in the price of a security associated to P-Space.

TAs are based on special distances between two objects in P-Space and they point out the more likely time for a major or intermediate reversal.

We do not use TAs in Quantum Trading to forecast at what price a top or a bottom could occur as we have done so far with QPLs, but rather for developing and studying the algorithms that can show us *when* a top or a bottom can happen, regardless of the price.

Using "Time and Price" Quantum Trading techniques at the same time, we'll be able to do very precise forecasting of trend reversals, as we see in the next chapter, which demonstrates how to use these instruments together. But now it is important to understand how to calculate "Time Algos" in Quantum Trading, and how to use them in the simplest way possible.

To discuss and analyze this subject in detail would take not only a few chapters, but an entire book, and probably more than one, because it is a vast subject that allows us to study in detail the expansion and the contraction of markets in different time frames, ranging from very long term to very short term.

PECULIARITIES OF TIME ALGORITHMS

W. D. Gann approached this subject by attempting to find a correlation between the geometrical angles created by the movement of different objects in the solar system and the movements of the financial markets. He often used the term *natural date* in his *How to Make Profits in Commodities* book. Some studies try to demonstrate that in the days when a top or a bottom occurred in certain markets, and Gann referred to the natural date, there also occurred a precise angular relationship between two or more objects of our solar system, or what is generally called a planetary aspect.

The approach we use in this chapter can be considered in some ways similar to the one used by W. D. Gann; however, we do not consider a cause and effect relationship between solar system objects and price movements, as Gann theorized.

On the contrary, I don't believe that planetary aspects can directly affect the stock exchange, commodity, or Forex prices, as Gann believed. I always refer to the time algorithm of Quantum Trading as affected by the angular relationships of objects moving, not in the solar system, but in P-Space, a mathematical, virtual space where objects follow rules of movement that are similar to those ruling celestial objects. These relationships generate different degrees of curvature in P-Space, thus creating the price-deflection phenomenon in which price behaves as if it were a light photon or as an electron jumping from one orbit to another.

These last two considerations dramatically change not only the approach, but also the results you obtain by approaching the markets using

Quantum Trading techniques instead of Gann's emphasis on planetary aspects.

Even though I spent many years researching and testing Gann's theory and ancient astronomical systems, my approach is inspired more by Einstein's concept of the curvature of time-space—or price-space, in my system—and the nonlocality approach of quantum physics with respect to the behavior of particles. Even if some of Gann's followers, without a thorough understanding of his work, could say that Gann also used what we call Quantum Price Lines, it is easy to demonstrate that the price-deflection concept stemming from the theory of relativity is totally absent from his work. Despite the fact that transforming longitude in price was an idea circulating among a small group of traders to which Gann belonged, the way he calculated his mysterious planetary lines is quite different, because he used whatever scale he considered good enough to fit the price movement, and not only in terms of the 2^n series. Instead, in Quantum Trading, we use a rigorous quanta package of units of price to calculate QPLs and their harmonics, to maintain the system in "equilibrium," something in which Gann did not seem to show a strong interest. The idea to transform celestial objects in longitude was present in Gann's work, but he never mentioned precise rules and left only a few observations in a couple of private letters, published many years after his death. Also, the way we calculate Gann Angles is very precise, using only the quantum conversion scale that is discussed in Chapter 10. We use this scale to calculate both Gann Angles and Quantum Price Lines, to maintain a "quantum approach" as we explain in Chapter 7.

Despite this, Gann's contributions can be seen as considerable and the direction he followed was totally different from the one we chose to develop our Quantum Trading models.

Regarding "Time Algos" in Quantum Trading, we can introduce the subject by considering that the curvature of the P-Space not only reflects price levels or the price deflection phenomenon, but that it also reveals the existence of a time-curvature that is acting upon and influencing the price *action.*

We can think of this time-curvature phenomenon as though in one particular moment of the continual time-flow in P-Space, we face a time "singularity" influencing the price behavior according to special conditions that have to be found in the angular distance between two objects in P-Space.

This special angular distance between two objects of P-Space creates a singularity in the time-space where the price is moving. This singularity generates another degree of curvature in whatever position the price is located. If it happens that simultaneously these, or other objects, generate through QPLs and QPLSHs an additional curvature at the level where the price is, then the P-Space is strongly warped and the price deflection phenomenon occurs. This Quantum Parousia appears to be generated by the simultaneous action of both price and time techniques, and shows us the conditions for a reversal in the trend, as we see more clearly in Chapter 14.

We call these phenomena the Time and Price Algos and the Price Deflection Phenomenon. The objects in P-Space can warp the space where the price of a security moves in two different ways, as indicated by QPLs and TAs:

- QPLs and QPLSHs warp P-Space around the price indicated by the P-Space Operator (PSO—see Chapter 5). The price indicated by PSO corresponds to a high degree of curvature where Einstein's light-photon deflection phenomenon is replaced by price deflection activity. The PSO identifies powerful price points of support and resistance corresponding to the P-Space curvature. QPLs and QPLSHs don't give you information about when the price of a security can reverse, but only at what price it can make a reversal.
- TAs are generated, like QPLs, by special angular distances between objects in P-Space. They can warp the P-Space at the point corresponding to today's price of a certain security, or the past or future price corresponding to the date of the TA. TAs don't give you information about at which price a security can find a strong resistance or support, but rather they tell you the time when this can occur.

If on a certain date the price of a security happens to be at the price level indicated by a QPL or QPLSH, indicating a curvature in P-Space, and simultaneously one or more TAs generated by slow-movement objects occur, indicating another P-Space curvature for today's price, then the price deflection action is very strong.

So we can see a reversal in the security trend because at the same time both price and time techniques warp the P-Space, generating the most powerful support or resistance possible. We call this Quantum Parousia.

A simpler way to explain TAs is to speak of them as if they were alternative clocks. If someone asks you the time, you usually take a look at your watch and answer, "It's 3:15 P.M." However, that answer takes too many things for granted. To be more precise, we should specify that it's 3:15 P.M. according to the Sun-Earth clock. If we then are asked what day it is, we can check the date on our watch or calendar and answer, "It's August 13th." But it's still the Sun-Earth calendar we are referring to because the calendar we use every day to organize our meetings, travels, and life is based on the revolution of the Earth around the Sun.

The question now is whether there is only this Sun-Earth clock or other types of clocks, such as Earth-Saturn, Earth-Mars, Mars-Uranus, or Mars-Jupiter, affecting our lives?

Certainly the ancient Egyptian astronomers were able to find a very high degree of correlation between a stellar clock, like the heliacal rising of Sirius and the flooding of the Nile River. It was necessary to forecast the annual flooding of the river because the fertility of the arable land stolen from the surrounding desert depended on the Egyptians' ability to be prepared to escape before the flood breached the banks of the river, depositing nutrients and minerals naturally contained in the sediment. Because the flood was very violent before the Aswan Dam was built in modern times, and the farmer sleeping along the riverbank could easily drown if the flood came at night, it was important to be able to forecast this phenomenon.

In this case we are speaking about the Sirius-Sun-Earth clock. TAs are about alternative clocks based on other celestial objects' special angular distances. The time for a reversal comes for a certain security when an "M" and "J" conjunction occurs, because the "M-J" clock tells us it is "noon" in its cycle. I don't know if this is relevant in human daily affairs, but it is definitely relevant in our P-Space to forecast the time of a reversal in financial securities.

The "M-J" clock is actually the "M-J" synodical cycle and there are correspondences between the hour on the clock face and the top and bottom of a certain security regulated by this specific "M-J" alternative clock in the P-Space.

The most challenging research is to identify for each security its particular alternative clock generated by one or more synodical cyles acting separately or together. This field of research is huge, but productive, because it provides the trader with useful information about the time when the trend of a security can change.

Price Families and Time Families in Quantum Trading

In the Quantum Trading model we consider the price of a security as one variable and time as another variable. Each variable originates a family of techniques: time family, which includes TAs, and price family, which includes QPLs and QPLSHs. We will open a new trade only when we simultaneously have an indication of a reversal provided by price family techniques *and* time family techniques.

Remember that P-Space is an interactive, virtual structure ruled by entanglement and nonlocality. Furthermore, P-Space is a multidimensional, virtual space composed of securities prices, time, and celestial objects in motion, which curve space-time due to their mass or gravitational effect.

Quantum Parousia occurs when we have a combination of trading signals coming from price family and time family techniques simultaneously. The price-time parousia, as we discuss in more detail in Chapter 12, is what we are waiting for, to open a new position according to our Quantum Trading model.

P-Space Object's Special Angular Distance

The objects of P-space can show different angular distances. Some special angular distances are more active than others and create a higher level of curvature in P-Space, and so a reversal or the acceleration of the trend of a security.

The most important special angular distances come from the circular dodeca-partite algorithm (CDPA), as introduced in Chapter 7.

Dividing a circle composed of 360 degrees by two and its multiples, we obtain:

$$360/2 \ = 180$$
$$360/4 \ = 90$$
$$360/8 \ = 45$$
$$360/16 = 22.5$$

Dividing the circle composed of 360 degrees by three and its multiples, we obtain:

$$360/3 \ = 120$$
$$360/6 \ = 60$$
$$360/12 = 30$$
$$360/24 = 15$$

Dividing the circle of 360 degrees by five and its multiples, we obtain:

$$360/5 \ = 72$$
$$360/10 = 36$$

The following are also active angles:

$$72 \times 2 = 144 \ \text{ and } \ 36 \times 3 = 108$$
$$45 \times 3 = 135 \ \text{ and } \ 60 + 90 = 150$$

We can calculate these angular distances both in heliocentric (helio) or in geocentric (geo) longitude, of course taking into consideration the helio position of object one with respect to the helio position of object two, as well as the geo position of object one with respect to the geo position of object two.

Where can we find the longitude of the P-Space objects to calculate their special angular distances? It is very easy. We can do it in the same way that we can find the longitude to calculate QPLs. We can simply buy the geo and helio ephemeris books, or use free programs on the Internet to calculate the longitude of objects in our solar system.

You simply check on a certain date what the longitude was for the various objects. Then you calculate the angular distance between the two objects and observe whether a special angle is forming. You have to do this by comparing the different angles between each pair of objects.

When two P-Space objects occupy the same degree and they have a 0 degree angular distance, they significantly curve the P-space because the mass of two objects sum each other. The result is a not only a price entelechy formed by two QPLs crossing each other but also a very strong Quantum Trading Time Algorithm (TA) that can signal a trend reversal or acceleration.

So far we have considered only the price aspect of an entelechy formed by the crossing of two different Quantum Price Lines or subharmonics (first QPL + second QPL, or a second QPLSH). But if the entelechy is formed by two QPLs, it implies that the first object and the second object always occupy the same degree on the complete 360-degree circle and have the same longitude, within -1 or $+1$ degrees.

The 0-degree angle can also be called a conjunction between two P-space objects.

An entelechy formed by two QPLs, plotted on your favorite security chart, can offer two main possibilities:

(1) The price is tangent to the entelechy. This is a special case in which we have a time and price signal at the same time, and it is one of the various kinds of entelechy (the first type) that we have seen in the previous chapters. It is the mother of all the various cases of what we have called a Quantum Parusia.

(2) The price of a security is not tangent to the entelechy formed by two QPLs, but is rather quite far away. In this case, the conjunction between the first object and the second object represents only a Time Algorithm, without giving any information of support or resistance price levels, as in the previous case. The time algorithms simply alert you to the possibility of a reversal or acceleration in the trend.

When two P-space Objects are connected at angles of 18, 90, 135, 45, 60, 120, 150, 108, or 72 degrees, they generate a powerful Quantum Trading Time Algorithm (TA).

The most powerful algorithm after the conjunction is the 180-degree angle. You will not find it in all security major tops and bottoms, but when it occurs it is a very important signal and should be carefully monitored, as it happened on the S&P 500 top in September 2000, just before one of the most severe corrections in the history of this stock index. The other angles obtained by CDPA, such as 22.5, 30, 15, 36, and 144 degrees, can also generate time algorithms.

The biggest angles between the slow movement objects occur very rarely, perhaps even decades apart. Other angles occur more frequently. Angles between "M" and other slow movement objects occur several times a year. For example, the time from a conjunction to a 180-degree angle of "J" and "S" objects is about 10 years.

Let's take a look at the TA hierarchy:

- "P," "N," "U," "S," "J," and "M" are slow-moving objects and show the major tops and bottoms. "M" is the fastest among the slow-moving objects and plays a key role.
- "Su," "V," and "Me" are fast-moving objects and show the minor reversals or help to indicate the date of a major reversal within the approximation of one or two days if they appear on the same day as an angular distance between two slow-moving objects.
- If more than one TA occurs simultaneously on a certain date TA, then the signal for a reversal or acceleration of the trend is stronger.

In TA Hierarchy we find on the first line the slow movement object series showing the major tops and bottoms, and the fast movement objects on the second line indicating the minor reversals, or helping to specify the day of a major reversal if occurring simultaneously with slow object TAs. Usually when we have a major reversal, both types of TA occur in the same day within a $+/-1$ range. For instance, we'll find a 45-degree angle between "J" and "U" objects, as well as a 72-degree angle between "V" and "S" objects on the same day.

Under the family of Quantum Time Algos we can mainly find the following TA techniques:

- Absolute synodical distances between two P-Space objects.
- Relative synodical distances between two P-Space objects from top and bottom.
- Sideral cycle of P-Space objects.

- Phases of P-Space objects.
- Singularity of a P-Space object movement.
- Special intraday Time Algos.

In this chapter and in this book, we'll study only some basic aspects of absolute sinodical distances between two P-Space objects because the subject of TAs is very large and it takes many books to be treated in a detailed way, reviewing all of the implications conveyed by the other TA techniques mentioned above. So, from now on when we refer to TAs we always mean absolute synodical distances.

FORECASTING TOPS AND BOTTOMS WITH TAs

Now we'll give you some examples of TAs that have forecast a reversal in different securities. We'll review the chart of some securities we have already studied using QPL techniques, but this time we'll use TAs. In this way, we'll discover how time curvature is crucial to forecast a reversal. We'll also review new charts to better understand the implications of time-family techniques.

In Figure 11.1, the top occurred at point A on July 15, 2008. Considering the geo longitude, we find a 36-degree angle between "J" and "N" objects.

FIGURE 11.1 TAs on EUR-USD Daily Chart

Simultaneously a 149.7-degree angle between "Su" and "N" occurs, as well as a 134.1-degree angle between "Me" and "N." The value of the 149.7-angle is very close to the important 150-degree angular distance that we take into consideration for a reversal, and the same is true for 134.1 degrees, which is very close to the very important 135-degree angle. Three days before the top occurred, a 0-degree angle between "M" and "S" occurred. As you remember, the conjunction is one of the most important aspects of a reversal.

Not all of the major tops and bottoms show a conjunction, but many do. In the same way, when a conjunction shows in P-Space, not all securities will make a top or a bottom, but those that at the same time show an entelechy generated by the crossing of two QPLs or a Quantum Parusia are very likely to have a reversal, or at least a significant acceleration of the trend.

At point B in Figure 11.1 on December 18, 2008, you can see an intermediate top on the EUR-USD chart. That day a helio 30-degree angle between "G" and "P" occurred, and also a powerful 90-degree angle between "M" and "U." TAs are very powerful and generate a singularity in P-Space. The price reacts immediately by dropping from 1.4615 to 1.2455, before reversing again.

In Figure 11.2, you can find an example of how Quantum Time Algorithms are able to forecast another major reversal in EUR-USD. At point A on December 3, 2009, the EUR-USD chart showed a major top and at the same time a helio 59.14-degree angle between "J" and "P" objects occurred, as well as a helio 108 degree angle between "M" and "U" objects. A

FIGURE 11.2   Other TAs on EUR-USD Daily Chart

FIGURE 11.3 GBP-USD Daily Chart Major Top with TA

59.14 angle is a value close to the very important 60-degree angle between two objects, and therefore is a red flag for a reversal.

If you consider, instead, the geo angular distances on the same day, there was a 45-degree angle between "M" and "S" objects and a 134.13-degree angle between "M" and "P" objects. Also, on the same day there occurred a 72-degree angle between "S" and "N" objects, and also a 30-degree angle between "V" and "P" objects. The day after the top on December 4, a 60-degree angle between "V" and "S" was shown.

This very powerful cluster between slow and fast movement objects in P-Space generated a major top from which the price dropped around 20 percent.

In Figure 11.3, we can see that on November 9, 2007, GBP made a top at point A against the U.S. dollar and a series of very important angular distances occurred simultaneously. A helio conjunction between "P" and "J," a 108-angle between "M" and "N," and a 59-degree angle between "U" and "Su" all happened on that day. The conjunction is the most important aspect that contributes to create singularity in the P-Space, which gives us a strong signal for a possible reversal. The other angles between the P-Space objects are also very strong. And therefore, they confirm that a top can occur in this time window. Of course, as we will show in detail in the next chapter, a time signal generated by a TA is not enough for us to open a "contrarian" trade in the opposite direction with respect to the trend. To do that we also need the confirmation that comes from a technique belonging to Quantum Price Family. If in the same time window indicated by a TA there is also an entelechy, or a QPL, or a QPLSH, or a Gann Angle that is

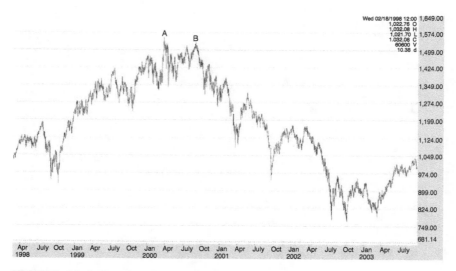

FIGURE 11.4 March 2000 S&P 500 Tops with TAs

touched by the price, then we can open a new trade, expecting a reversal. If we have already opened a position a few days or a week before, we can close this position when the parusia occurs.

Chapter 10 discusses how QPLs and QPLSHs can identify very strong levels for a reversal in the S&P 500. In Figure 11.4 you will appreciate how Quantum Time Algorithms can point out the time when the most important tops and bottoms in the S&P 500 occurred.

On March 24, 2000, the S&P 500 Index made its historical top and on the same day many important TAs occurred, such as helio 9.1-degree angle between "S" and "U." Also, on the same day, geo longitude shows very important angles of 91 degrees between "J" and "U" and 60 degrees between "J" and "V" closed.

At point B, we have a double top on September 1, 2000, and many very important angles trigger a singularity and a high degree of curvature in P-Space. In fact, that day we have a geo 179.8-degree angle between "J" and "P" and a 90-degree angle between "J" and "Su." Simultaneously we see three very important angles, such as a 108-degree angle between "Me" and "S," a 149-degree angle between "Me" and "U," and a 135-degree angle between "Me" and "N."

As we mentioned at the beginning of this chapter, after the conjunction, the 180-degree relationship is one of the most active angles to predict a reversal.

The 2007 S&P 500 historical top that we have already studied using QPLs was also triggered by very important angles between our P-Space

FIGURE 11.5 S&P 500 Index Major Top and Bottom with TAs

objects, or TAs, as you can see in Figure 11.5. On October 11, 2007 (point A), we find a helio 149-degree angle between "M" and "J," and a 90-degree angle between "N" and "V." The same day, other very important geo angles occurred, such as an 89.6-degree angle between "J" and "U" and a 59.4-degree angle between "S" and "M." Consider that just two days before the actual top, the 90-degree angle between "J" and "U" was perfect, and that it takes a little less than three years to have a new 90-degree angle between "J" and "U."

Considering the helio "J-U" cycle, after the conjunction, the 90-degree angle shows after about three and a half years, and after another three and a half years the 180-degree angle closes. After three and a half years, the other 90-degree angle shows up, and after another three and a half years the conjunction occurs again. The helio "J-U" cycle completes and repeats about every 14 years. This approach is only superficially similar to the studies of cycles that are at play in various securities, because the latter takes only linear time into consideration, and tries to find a correspondence between fixed time cycles repeating themselves from the past to the future. Instead, our TAs are based on the elliptical nature of time, because the time between one 90-degree angle and the following one is not constant and can vary by several months, being ruled mathematically by the Kepler equations showing that planets sweep out equal distances in different time frames, which is a direct consequence of the law of areas, according to a line that connects a planet to the sun sweeping out equal areas in equal time frames.

So the 90-degree angle between "J" and "U" is very important, and when it happens it should be carefully considered as a trigger for a reversal in many securities, if a Quantum Parusia also shows at the same time. We were able to reach the top of the S&P 500 in 2007 because time-family techniques and price-family techniques simultaneously provided strong signals for a reversal, and the parusia showed up in the S&P 500 chart, as the next chapter demonstrates.

In Figure 11.5, at point B, the S&P 500 finally reversed after one of the biggest drops in its history, on March 6, 2008. On this day, you can find a geo "M" and "N" conjunction (0-degree angle) and also an "N" and "Me" conjunction. This double conjunction occurred on the same day and is a powerful signal for a reversal that in fact took place that exact day. The following day, on the weekend, "Su" and "S" showed a 180-degree angle and a 60-degree angle between "J" and "V," and it is another signal adding to the previous cluster.

If we study the helio angle we find, still at point B, a powerful conjunction between "M" and "J," and simultaneously a 45-degree angle between "M" and "U."

At point A in Figure 11.6 we find powerful TAs that can be very useful to forecast a reversal if confirmed by price-family techniques. On March 17, 2008, Gold Comex future topped at 1021 with a Helio 44-degree angle between "J" and "N," a 144.6-degree angle between "N" and "Me," and a 150-degree angle between "S" and "Su." On the same day, there were very

FIGURE 11.6 Gold Daily Chart Major Top and Bottom Using TAs

important geo angular distances, such as a 60-degree angle between "J" and "U," and a 119-degree angle between "M" and "Su," as well as a 121-degree angle between "M" and "V."

At point B in Figure 11.6, we find the bottom of November 13, 2008. There is a very strong geo angle of 179.09 degrees between "S" and "U," a 119.15-degree angle between "S" and "J," and another 60-degree angle between "J" and "U." The day before, a powerful conjunction between "Su" and "P" confirmed that the trend of gold was ready for a reversal after a drop of about 15 percent.

In Figure 11.7, we can see one of the biggest bottoms on USD-JPY, which occurred on April 19, 1995. Taking a look at the geo aspect, we'll find a 60-degree angle between "P" and "U," a 149-degree angle between "P" and "Me," a 61-degree angle between "N" and "Su," and a 45-degree angle between "J" and "U." The day after, we'll also find a 90-degree angle between "Me" and "U," a 45-degree angle between "V" and "S," and a 135-degree angle between "J" and "M." This is a big cluster of both angular distances formed between the very slow objects, as well as between the high-velocity objects in P-Space. Review the TA hierarchy discussed earlier in this chapter for an explanation of slower and faster moving objects.

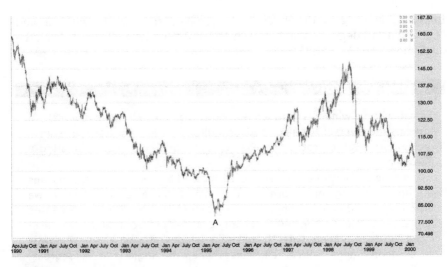

FIGURE 11.7   USD-JPY Daily Chart

TIME ALGORITHMS' GOLDEN RULES

To forecast a reversal we have to follow these three steps:

1. Examine if a special angular distance occurred between two slow objects. Because the time window can remain open for a few weeks, we need another indication of other TAs to be more precise within the shorter time window of a few days. To find this shorter time window, we proceed with the steps below.
2. Check to see if we have special angles being formed between a slow-moving object and a fast-moving object.
3. Check to see if we have special angles being formed between fast-moving objects.

    If there is a special angle between two slow-moving objects and, at the same time, one of these two objects forms another special angle with a fast-moving object, this is a strong combination that allows us to forecast a reversal or a trend acceleration within plus or minus 1 degree.

Let's continue with the USD-JPY bottom in Figure 11.7. A few days after the bottom on April 26, 1995, there were a long series of special helio angular distances. "P" and "S" showed a 107.64-degree angle (very close to 108 degrees), "P" and "J" showed a 45-degree angle, and "J" and "M" a

72-degree angle. Furthermore, there were angles between slow- and fast-movement objects, such as a 72-degree angle between "J" and "Su" and a 150-degree angle between "J" and "V."

In this example, the golden rule was verified. Simultaneously, we have a TA formed by two slow-moving objects, such as a 45-degree angle between "P" and "J," and another TA formed by the same slow objects with one or more fast-moving objects. In fact, "J" forms not only a special angular distance with "P," but also with "M," "Su," and "V."

This is one of the most powerful patterns we have discovered in TAs.

There are many other powerful patterns in TAs that show relationships between many different angles, but this would take an entire book dedicated solely to TAs, while this book is based on identifying QPLs and QPLSHs. In our seven-day, full-immersion classes in Europe, America, and Asia, we study in detail the most effective TA combinations and how a bottom can be forecasted by studying the TA that occurred in the previous top and/or bottom.

There are many rules that show you how to select the right chain of angular distances to forecast a reversal very precisely with TAs. This is one of the most fascinating parts of Quantum Trading techniques, and by studying this subject continuously, we discover different patterns of TAs.

In our classes we explain in detail the application of TAs in intraday trading, and students learn to apply them to their daily trading during the postcourse tutoring period that we offer with our classes.

CHAPTER 12

How to Use All the Trading Tools Together

As seen in the previous chapter, there are many Quantum Trading tools such as TAs, QPLs, QPLSHs, and Entelechy. In real trading, we use all these tools together to obtain as many signal clusters as possible.

As you learned in the previous chapters, we have two families of Quantum Trading techniques: the price family and the time family. I don't usually trade the markets opening a new trade relying on only one family of techniques, but instead prefer to wait until different tools from Quantum Trading theory, stemming from both families, give the same signal to buy or to sell.

Coming back to Einstein's theory of relativity, it is very important to remember that P-Space is very similar to the space-time-universe conceived by Einstein. In our P-Space, the space-time Einstein binomium is substituted by the price-time binomium.

Price-time in our Quantum Trading model is everything. To get a full signal forecasting a reversal we need both the price techniques and time techniques to confirm simultaneously the buy or sell signal. This is what we call "Quantum Parousia."

QUANTUM PAROUSIA

The combination of a price-deflection caused by the presence of a mass in P-Space, as we have studied so far with QPLs, and the simultaneous singularity in Time (TAs) generates the most powerful condition for a reversal, where time and price generate the highest level of curvature possible in P-Space. This is what we call Quantum Parousia, or simply parousia. Quantum Parousia occurs when we have a combination of trading signals coming from price-family and time-family techniques simultaneously.

Parousia is an ancient Greek word meaning a theophany of God, or "return." We borrow this term from the ancient Christian religion because *Quantum Parousia* elegantly describes the cyclical "return" of an event considered the highest manifestation of its potentiality, in our Quantum Trading model, exactly as the term *parousia* implied for the ancient mystics the beauty and absolute perfection of the highest and most powerful mind in the universe manifested in sensible form.

In our Quantum Trading model, parousia is something similar, but more important than Entelechy. Entelechy expresses the concept of a perfect condition in a dynamic process, but parousia alludes to a higher order of perfection, the highest you can find in our P-Space universe, where time and price are in equilibrium.

In Chapter 11 we review the time Quantum Techniques that give us an understanding of the most important time windows for a reversal. These time windows can last several weeks for big trend reversals. Using other time-quantum techniques we can obtain time windows for a reversal lasting only two days, and with other intraday Quantum Trading techniques we can get even smaller time windows lasting only a few hours to forecast a reversal. Using these three types of time windows together we are able to be more precise in forecasting a change in the trend and open very profitable trades.

Returning to our price-time trading approach, or parousia, I want to stress that it is very important to open a new trade only when we have at least two time signals and one price signal at the same time.

When time squares the price, the trend is ready for a change and a reversal can happen, according to one of W. D. Gann's ideas. In Quantum Trading, we don't speak in terms of time squaring the price, but rather in terms of the time when the price-photon-particle of light is likely to meet the highest level of curvature in P-Space. So, the presence of objects in P-Space fundamentally curves the space, warping it in the same way that the weight of an iron ball will warp a tablecloth that is held in the air. The degree of P-Space curvature can be higher than in other moments, depending upon the relationship between the objects and the different special distances between them, generating TAs.

If the price touches a QPL or a QPLSH, and in the same window of time one or more TAs occur, then a reversal is very likely to happen because we have a "Quantum Parousia."

Will a reversal always occur when we find such an equivalent of price and time in the P-Space? The answer is no! About 70 percent of the time it will occur, but in the remaining 30 percent of occurrences, it won't happen.

However, during the 30 percent of the time it does not occur, we can make even more money, because when the price breaks the QPL instead of changing its trend, we very often see a steady continuation of the trend in the same direction. The price will accelerate its speed, showing a very fast run-up or a dramatic plunge.

The first thing we have to do is display on our favorite chart all of the QPLs and QPLSHs that can be more relevant for that section of the trend. It is not difficult if you have advanced software that can display all of the data at the same time.

Second, we have to try to understand which QPL or QPLSH matches the previous top or bottom.

Then we have to check if the QPL or QPLSH you find close to today's price is generated by the same object that generated the QPL or QPLSH that you found on the previous top or bottom. In case today's price is far from a QPL or QPLSH, we wait until this price level will be reached. Then we have to immediately check if an entelechy, formed by two QPLs or QPLSHs, is forming in the proximity of today's price. In this way we try to spot if an entelechy is close to occurring in the next few days or weeks.

If the price touches the entelechy it will be a very good entry signal. Remember that we have three different kinds of entelechy. So we can also put Gann Angles on the chart at the same time. Sometimes you will find amazing signals indicated by an entelechy composed of a Gann Angle and a QPL. But this will not always happen. Other times, we'll find on the chart entelechies composed of two quantum price lines. Other times, even though you don't spot any entelechy on the chart, you will have a big reversal signal composed by a quantum time technique and a quantum price technique, a parousia. This is a very strong and important signal, and when it appears, carefully follow the price dynamic to be ready to trade the market for a reversal.

A Precious Rule of Thumb for Real Trading

As a rule of thumb, when the price is rising and it meets a QPL or QPLSH, if there are no time signals, the price will usually continue following the same trend without any change or reversal. Instead, when the price touches a QPL and at that very moment we also have a time signal, seven out of 10 times a reversal will occur. This is a new concept developed in Quantum Trading theory and proven true in the past 10 years of trading. There is nothing similar to this in other theories such as the Gann theory, or other approaches that in some way can be considered similar to this proprietary approach.

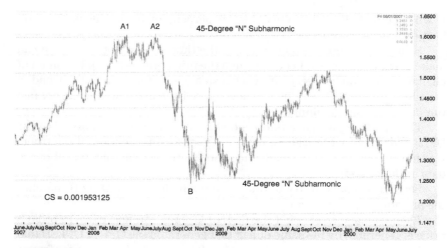

FIGURE 12.1 EUR-USD Daily Chart and Parousia

As we have seen in Chapter 4, there are different ways to study how solar storms can affect the stock markets. Other traders have tried to find a correlation between the different planetary angles and the increase of solar-storm activity and their consequences affecting the financial markets. Other researchers try to find other correlations between planetary angles and long-term and medium-term cycles in stocks' and commodities' price behavior. With respect to Quantum Trading, these kinds of approaches can be considered only partial and incomplete, not relying on a price-time equilibrium approach.

Now we'll study together many examples of Quantum Parousia.

In Figure 12.1, the top occurred at point A2 on July 15, 2008, and we find here a Quantum Parousia.

In fact, on this day we find a perfect match of quantum price-family and quantum time-family techniques. There are two 45-degree "N" subharmonics (QPLSHs) offering a very strong resistance that could arrest the uptrend. This indication comes from a price-family technique, as we study in Chapter 11.

If you review the set of special TAs that was shown in Chapter 11 for this top on the EUR-USD daily chart, you can find many algorithms coming from price-family techniques on the same day. We have a geo 36-degree angle between "J" and "N" objects. At the same time, a 149.7-degree angle between "Su" and "N" occurs, and we also see a 134.1-degree angle between "Me" and "N." The value of the 149.7 angle is very close to the important 150-degree angle that we take into consideration for a reversal, and the same is true for the 134.1 angle, which is very close to the very important 135-degree angle. But it's not finished here. Three days before the top occurred, a 0-degree angle between "M" and "S" appeared. This conjunction is one of the most important aspects of a reversal.

So, we can open a short position on EUR-USD on July 15, 2008, because we have noticed a very strong parousia in P-Space and the curvature degree is very high. All the conditions we need to forecast a reversal are on the chart at the same time. On this day the EUR topped on USD and a severe downtrend began.

At point B, still in Figure 12.1, you find the bottom of EUR against USD on October 27, 2008. The TAs present at this time were a helio 89.9-degree angle (that is, practically 90 degrees) between "N" and "M." Taking a look at the geo aspect, we find a 107-degree angle between "Su" and "N" and a 179-degree angle between "S" and "U"—which is a very strong special TA, because 180 degrees is one of the strongest algorithms for a reversal, or to forecast the end of a trend, and a 60-degree angle between "J" and "M."

At point B we also find another 45-degree "N" QPLSH, originating from the same object that generated the QPLSH just on the previous historical top from which the downtrend started.

So we have a complete set of both time techniques and price techniques, a parousia that allows us to eventually close the short position and open a new long position at point B.

Please notice that even though, after point B, we have a double bottom, nonetheless, the trend dramatically accelerates until it reaches point C in Figure 12.2.

FIGURE 12.2 EUR-USD Daily Chart and Spike Top Parousia

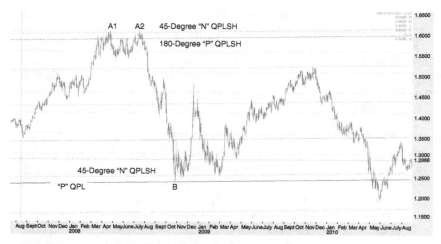

FIGURE 12.3 EUR-USD Daily Chart and Another Parousia

At point C in Figure 12.2 on December 18, 2008, you can see another Quantum Parousia and the spike top on the EUR-USD chart can be forecasted, thanks to the following complete set of signals coming both from price and time family techniques. As for the time techniques, there is a helio 30-degree angle between "G" and "P" that occurred, along with a powerful 90-degree TA between "M" and "U." These TAs generate a large singularity in P-Space. The price reacts immediately by dropping from 1.4615 to 1.2455, before reversing again because it meets, at point C, a 22.5-degree "N" QPLSH that is a resistance.

The parousia is perfect and the trend reverses at point C.

Besides the 45-degree "N" QPLSH (Figure 12.3), you can also consider the other very powerful 180-degree "P" QPLSH at points A1 and A2. At point A1 on July 15, 2008, you will find not only the TAs that we have already studied, but also the 180-degree "P" QPLSH. We can consider this kind of parousia even stronger than before because a cluster of QPLSHs generates a price signal. It is also very interesting for another reason. At point B, we find a price-family signal for a reversal: the "P" QPL, that is, the strongest level of resistance/support you can find in quantum price lines, because it is the origin of all the subharmonics (QPLSHs), and so it is the most important between whatever QPLSH, like the 45-degree "N" QPLSH, we can find near the bottom at point B.

So, what you have seen so far should encourage you to spend time on studying Quantum Trading techniques to be more effective in your trading. It takes time to be acquainted with the arsenal of all Quantum Trading tools, but it's worthwhile.

FIGURE 12.4 NASDAQ Futures Daily Chart Showing "P" QPLs (CS = 8)

Despite the fact that Quantum Trading offers you an autonomous system to trade the financial markets without referring to contemporary technical analysis based on oscillator and indicators, you can also use these very popular tools if you wish, to filter the signals obtained by QPLs and TAs. The professional trading system we use to manage our client's assets is based on Quantum Trading algorithms (algos). Basically, we use QPLSH and TA, but there are many other algos for short- and very short-term trading that can be learned at another time. If you invest time in studying our techniques, you will surely be rewarded by excellent results.

In Figure 12.4, you can again see the NASDAQ daily chart that is shown in Chapter 10. Studying the TAs we find at point A, on March 24, 2000, a helio "N"–"Su" 120-degree angle, a 91-degree angle between "U" and "S," and a 91-degree angular distance between "U" and "M." Furthermore, we also have a conjunction between "M" and "U" on the same day, and another 135-degree angle between "M" and "Su."

Let's review the price family indicators for this top. At point A, we find a "P" QPL exactly on the price of the same-day historical NASDAQ top. We remind the reader that we are using a CS = 8. The price is tangent to the "P" Quantum Price Line, and when it touches the line, it suddenly reverses.

Because we receive very strong signals from both the price family and the time family, we decided that it is worthwhile to open a new trade, selling short the future NASDAQ. Probably everyone will think that you are absolutely crazy if you tell them that you are selling short the NASDAQ future if the price keeps rising like that, but you have found a very strong parousia according to our Quantum Trading rules. The parousia is formed on the price side by the "P" QPL, that is, the strongest line for a top. For this reason we can use this very strong parousia to observe the behavior of the price in the proximity of this very high level of P-Space curvature. Because the price starts to react as it touches the QPL reversing its trend, we can open the short selling. Whether or not opening the trade just as the price touches the QPL or waiting for some days to receive another confirmation from the quantum price dynamic, such as a series of consecutive 60-minute or daily bars with lower lows, it depends on how aggressive you are in your trading. Of course, even if you consider yourself very aggressive, you can always start by selling only one third of the total position that you intended to sell on the top. So, if you wanted to start with a three NASDAQ future size, you can open just on the "P" QPL price an initial position price of one contract only. You can sell another contract tomorrow if the price breaks today's daily bar low. You can sell at another NASDAQ future in two days if the price breaks tomorrow's daily bar low.

In Figure 12.5, we still have the NASDAQ future top (point A) that we have studied in the previous section, but this time we are using a new set of QPLs with CS = 16 instead of CS = 8, and the same TAs we have just studied. The parousia forms even when we use a different QPL scale, and this is another indication of a very strong reversal.

In Figure 12.6, we have again the NASDAQ future chart. This time we are using 90 QPLSH and a CS = 8. At point B on September 1, 2000, we find another parousia because we receive signals for a reversal of the trend from both price and time families of Quantum Trading techniques. Studying the TAs of this day, we find a geo 180-degree angle between "P" and "J" and a 90-degree angle between "P" and "Su." We also find a 135-degree angle

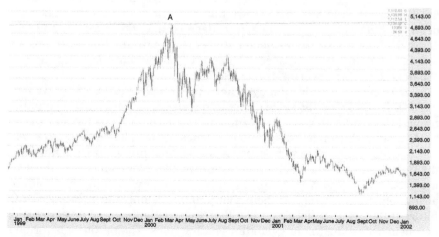

FIGURE 12.5 NASDAQ Futures Daily Chart Showing "P" QPLs (CS = 16)

FIGURE 12.6 NASDAQ Futures Daily Chart Showing "P" QPLs and 90-Degree QPLSHs (CS = 8)

between "N" and "Me" and a 150-degree angle between "U" and "Me." There is also a 90-degree angle between "J" and "Su."

On the price side we can find a very strong resistance at point B, indicated by the "P" 90-degree QPLSH. The parousia is perfect and so the trend plunges after a pullback because it finds a strong resistance after having tried to pull back to reach the top again.

Finally, in Figure 12.7, we find, still on the NASDAQ future daily chart, with the same initial CS = 8, but this time we have drawn the "P" 72-degree

FIGURE 12.7 NASDAQ Futures Daily Chart Showing "P" QPLs and 72-Degree QPLSHs (CS = 8)

FIGURE 12.8 NIMEX Crude Oil Future (CS = 0.5)

QPLSH. It is very interesting to notice that at point D this passes a QPL that arrested the severe NASDAQ plunge, which began at point C, and offered a very strong support on the bottom that occurred on November 21, 2008 at 1016.

On the time side we find the following TAs. We have a geo 120-degree angle between "J" and "S," a very strong TA. Then we have a 46-degree angle between "M" and "J."

This is another example of Quantum Parousia, and the trend reverses as expected, according to our Quantum Trading model.

The NIMEX crude-oil chart gives us another chance to verify how the historical top in 2008 occurred on a Quantum Parousia. As we see in Chapter 12, there is not only a very strong "J" QPL passing on the same top price level (see Figure 12.8, point B). We also have an entelechy on the top on July 15, 2008, formed by the crossing of "J" and "Su" QPLs (see Figure 12.9, point A) that before studying TAs could be considered sufficient to forecast a change in the trend and open a medium-term short position on crude oil. Now we'll try to discover together if besides the entelechy, we also have TAs that can strengthen our belief that the trend is really changing.

We can find a 45-degree angle between "M" and "Su," a 36-degree angle between "J" and "N," a 45-degree angle between "S" and "Su," and a 150-degree angle between "N" and "Su." It's really a big cluster of Time Algorithms all occurring on the same day.

There is also a geo conjunction between "M" and "S" just four days before the top, when the price was on the same level as July 15. The conjunction is very interesting to analyze not only because, as we already observed many times it is the most powerful TA to be monitored

FIGURE 12.9 NIMEX Crude-Oil Future (CS = 0.5)

for a reversal, but also because normally when it happens, then two QPLs always cross each other creating the potential for an entelechy of the first type. But to have a true entelechy the price has to be exactly at the intersection of the two QPLs, and it doesn't happen all the time. In fact at other times, as in this case, the price is far from the intersection of the two QPLs. The intersection of "M" and "S" QPLs (see Figure 12.10, point A) on our crude-oil future chart that day was at 77.6 dollars, while the price was at 146 dollars. A very great distance! In this case the crossing of two QPLs doesn't form an entelechy, but just a

FIGURE 12.10 NIMEX Crude-Oil Future (CS = 0.5)

very strong TA, that is, a time algorithm announcing a potential reversal. In this case, to be more confident about a changing trend and the opening of a new trade in the opposite direction of the previous trend, we need to find a confirmation coming from another QPL passing at 146 dollars. If we find it right there, then we'll have at the same time a price and a time signal. For this reason, we consider that the equilibrium of Quantum Parousia can manifest in the P-Space, and so we actually open a new trade.

In our case we find the helio "J" QPL just on the top of crude oil. For this reason our parousia is considered perfect, and the crude-oil trend changes from bull to bear.

Not All Intersections between QPLs Are Entelechies

We can find on a certain date the intersection of two QPLs. When they cross each other, it is because there is always a conjunction in P-Space of the objects generating the intersection of two QPLs.

If it happens that the price of a security is just at the intersection point, then we have an entelechy. In the case that the price is quite far from the crossing of two QPLs, the intersection does not create an entelechy, but just a TA—that is, a time signal corresponding to a conjunction, the strongest between all the various special angular distances generating TAs.

In Figure 12.11, we see that at point A, a severe bear trend starts ending at point C, where a 120-degree "P" QPLSH passes just at the price level

FIGURE 12.11 NIMEX Crude Oil Future (CS = 0.5)

of February 12, 2009 (33.40 dollars), offering a very strong support. In fact, reviewing the rules we have observed many times, we try to forecast the level where the price will find a support to reverse its trend on a subharmonic of the QPL that was on the previous top, if we don't have another QPL generated by the same object in P-Space.

On February 12, we find many TAs: a geo 36-degree angle between "P" and "J," a 179.15-degree angle between "U" and "S" (that is very close to 180), and another 45-degree angle between "M" and "U." On top of that, a few days later, a geo conjunction occurred between "M" and "J," a very strong TA confirming that it might really be the time for a reversal.

Please notice that a geo conjunction between "M" and "S" marked the previous top of crude oil from which this big downtrend originated, and on the bottom from which the trend changes again we find another conjunction, this time between "M" and "J." Also, this time the conjunction is not an entelechy because the price is very far from the intersection of the two QPLs, but nonetheless the "M" and "J" conjunction is a very strong time algorithm that allows us to forecast a change in the trend. Furthermore, it is interesting to note that "M" is the pivot around which the "J" and "S" synodical cycles mark the change of the trend for crude oil in P-Space.

For this reason the picture is now complete, and we can expect, based on our Quantum Trading techniques, that at around February 12, 2009, time and price are now mature for a change in the medium term. The Quantum Parousia is perfect and we can close our short position on crude oil and open a new, long, medium-term position.

Let's study now the CMX silver future chart (Figure 12.12).

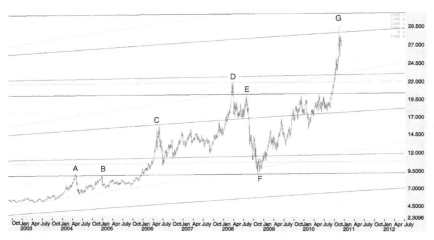

FIGURE 12.12 CMX Silver Future Daily Chart (CS = 0.03125)

We have many tops and bottoms to review to verify if we can find a Quantum Parousia that, according to our model, conveys all the conditions we require to forecast a change in the trend and open a new trade in the direction of the new trend.

On April 2, 2004, we find at point A various TAs such as a helio 150-degree angle between "N" and "J," as well as a 120-degree angle between "N" and "Su." We have a 90-degree angle between "S" and "Su" and a 45-degree angle between "S" and "Me." There is also a very strong and powerful conjunction between "M" and "S" a few days after the top, and the price is more or less at the same level as point A.

At point A we find a powerful resistance generated by "P" QPL. We have signals coming from both price and time families, and so we have a parousia.

At point B on December 2, 2004, we have a double top and a parousia with a "P" QPL offering a resistance on the price side. Coming to the Time Algorithms (TAs), we have a helio 150-degree special angle between "P" and "S," a 135-degree angle between "U" and "S," and a 150-degree angle between "U" and "J." We also have a 72-degree angle between "S" and "J," as well as a conjunction between "J" and "V." Wow! What a big cluster of TAs.

The Quantum Parousia is perfect and we see from point B a drop of more than 20 percent. You probably cannot easily realize that the drop is that dramatic by looking at Figure 12.12; because of the subsequent large price movement, it is difficult to understand the magnitude of the drop at first sight.

In fact, after a while, the trend reverses and a big bull campaign begins, ending at point C. On May 11, 2006, the bull campaign is arrested by "S" QPL offering a very strong resistance that silver fails to break.

It's amazing, considering how this resistance cannot be forecasted by traditional technical analysis tools, how precise Quantum Trading is in pointing out the most crucial points for a trend change. If we devote enough time to becoming acquainted with these techniques, with this information you can really make a lot of money. The TAs provide us with precise information. We have a helio 150-degree angle between "S" and "U," and a 150-degree angle between "J" and "Me." On the geo side, we find a 121-degree angle between "J" and "U," and another 121-degree special angle between "J" and "M."

Please remember the rules explained before about the importance of one object forming two different angles at the same time with two other objects. This is what we call "TNT," a very explosive combination.

Considering that we have signals coming simultaneously from quantum price techniques and family techniques; then we have a parousia, and we can trade it.

At point E, on the price side the resistance is offered by the "P" QPL. On the time side, instead, we have a geo conjunction between "M" and "S," and a 36-degree angle between "J" and "N." Furthermore, we also have an "Su" and "M" 45-degree angle as well as a "Su" and "U" 120-degree angle, plus a 150-degree angle between "Su" and "N." The pivot rule is verified once again, because "Su" forms three special angles with different objects in the same day. Considering the very strong effect of the conjunction, the curvature of our P-Space shows a very high magnitude and the parousia is just there, where a dramatic change in the trend occurs.

After a while the silver price reaches the inferior "P" QPL at point F, finding a strong support. It's amazing because the previous top at point E was made just on the higher "P" QPL. The TAs are a geo 180-degree special angle between "S" and "U," and also a 59.6-degree (very close to 60-degree) angle between "M" and "J." The parousia also manifests itself, and we can open a long trade taking advantage of this very big bull campaign.

Let's now study the parousia on Euro stock 50 index (Figure 12.13), the biggest European top-stock index. On March 7, 2000 (point A), there were many TAs, such as a helio 151-degree angle between "P" and "J," and another 141-degree angle between "P" and "M" (the day after). Still, on March 7, 2000, we also find a 90-degree special angle between "S" and "U," a 120-degree angle between "S" and "Su," and a 120-degree angle between "S" and "V." Please note that "P" forms two special angles simultaneously with "J" and "M," while "S" forms three special angles with "U," "Su," and "V." This is what we call an extraordinary cluster of special angular distances

FIGURE 12.13 DJ ESTOXX50 Index and Parousia

FIGURE 12.14 DJ ESTOXX50 Index and Another Parousia

generating very powerful TAs. Don't ever forget the rule we introduced previously—that when an object forms simultaneously occurring special angles with at least two other objects, it is very important. If this happens between two very slow-moving objects, we have a time window of just a few weeks in which the reversal could occur; but if at the same time it happens that there is more than one special angle between one slow-moving object and a faster-moving one, such as "V," "M," "Su," or "Me," in this case the time window narrows to one or two days only. When this happens, the possibility to catch a very big trend at its beginning is very high. In this way, you can really make a lot of money.

Coming to the price side of the Quantum Parousia, we find on the top the "N" QPL. So, we have signals coming simultaneously from both price and time families that the index is ready for a reversal, as actually happened.

In Figure 12.14, you can see that the index found a very strong support on both "N" and "U" QPLs at point B on September 21, 2001. That day a helio 121-degree angle between "N" and "S" occurred, and at the same time there was a very powerful conjunction between "M" and "U." From that day onward, the market rallied for around four months, recovering strongly and bouncing back.

Notice that the Quantum Parousia on the bottom of September 2001 is perfectly formed: After having reviewed so many examples, the reader can have strong confidence in Quantum Trading techniques and the parousia model.

CHAPTER 13

How to Use Options

Unlimited Gains with Little Risk

Options are extraordinary tools for gaining unlimited profits with little risk, especially if you use them together with Quantum Trading tools. In this way you can create effective strategies to protect yourself from future risk, as well as profit when stocks, or another security, rise or fall.

Before learning how to use options to trade the markets according to QPLs trading signals, it's better to review some basic concepts about the components and the strategies that we can build up with options. Then we'll be able to successfully apply these financial instruments to Quantum Trading to get the best result.

OPTION

An option is a contract between two parties, where one pays the other a premium to have the right, but not the duty, to buy or sell at a fixed price and obtain in the future a certain performance from the other party.

OPTIONS DEFINED

Let's begin at the beginning. The history of the option traces back to ancient Greece. Thales of Miletus, a Greek astronomer and philosopher, was the first investor who is known to have speculated, and he was able to earn a high income from options. It seems he was not only a great philosopher, but also an extraordinary businessman.

Aristotle tells us the story of how Thales, since it was still winter, forecast through astronomical observations the meteorological auspices for the upcoming olive harvest. With a relatively small amount of money, Thales approached the owners of the olive-presses of the area and offered them a cash deposit (today we call this an option premium) to gain the exclusive right to their presses. The crop was so abundant that Thales earned a fortune by reselling his exclusive rights to the olive presses.

Note that Thales probably could not afford to buy all the olive presses: instead, he limited himself to buying the right to use them within a certain expiration date, and he then resold the same right at a higher price, profiting on his investment.

Options later became widespread in the seventeenth century in different European countries to cover the risk of price oscillation in agricultural markets. Charles Mackay documents how in 1634 there was a wide and fluid use of options in the tulip markets. Tulips were considered a status symbol for the wealthy at that time, and their popularity skyrocketed amongst the aristocracy and bourgeoisie all over Europe. So, because of both speculation and high demand, the price of tulips soared in a period of feverish trading called tulipomania.

People habitually sold tulips for future delivery. Wholesalers preferred to stipulate an options contract with the horticulturalist to protect themselves from price oscillations. The wholesaler paid a premium to fix the price of the tulips and in doing so protected himself from price fluctuations. Usually he was able to make a profit and he actually bought the right, but not the duty, to buy the goods at the prefixed price. So, because the prices were continuously rising, the wholesaler paid for the tulips at a price set by the horticulturalist—which is similar to the option strike price—and then resold the right to the tulips to a client at a higher price for an extra profit. If the price started to fall, the wholesaler was free to buy from anyone else at the current price, renouncing the right to have the goods delivered at the prefixed, higher price.

At a certain point in 1638, tulip fever came to an end. Tulip prices, after soaring for a long time, collapsed, and the wholesalers who had used an options contract were safe from the ruin. On the contrary, the other traders who bought and sold with cash were ruined by the collapse of tulip prices. It was one of the worst speculative bubbles in history.

The listed options that you can buy today work similarly, and they are regulated by simple and clear rules for every trader. Listed options are tools with which you can carry out transparent, fast, and reliable transactions. That's why they always continue to become more popular.

At the beginning of the twentieth century, the options markets on stocks was already rich and lively, even though it was just an OTC market, which means it wasn't regulated by an options exchange. Brokers offered their clients the possibility to buy and sell options, but unlike today, those options showed a very strange pricing. Black and Scholes hadn't yet discovered their option pricing models (they weren't even born yet), and so, in the beginning there was a fixed price for both calls and puts. It wasn't important that 15 or 30 days were left until the expiration day of a period: the price was always the same. Option buyers and sellers didn't have any models to understand if an option price was fair or not. They had only a fixed price, which caused fewer problems than the ones experienced by a professional options trader today.

Finally, in 1973, a stock exchange was created in Chicago where many stock options were listed. Prices were clear and certain, and the transactions were regulated. But at that early stage only call options were listed. Put options weren't available yet; traders needed to wait a few years before having the complete option system that we have today. At that time traders used synthetic positions to simulate a put, obtained by selling short the stock, and simultaneously purchasing a call.

Options Characteristics

You can create an option on whatever is exchanged between different counterparts. If the option's underlying is listed in a regulated market, then the option is a standard contract with an official quotation. You have two kinds of options: calls and puts. A call buyer purchases the right to take advantage of a rise in the underlying. A put buyer bets on the decline of the option's underlying price.

Consequently we will review the definition of an option we already saw at the beginning of the chapter.

An equity option is a contract that conveys to its holder the right, but not the obligation, to buy (in the case of a call) or sell (in the case of a put) shares of the underlying security at a specified price (the strike price), on or before a given date (expiration day). After this given date, the option ceases to exist. The seller of an option is, in turn, obligated to sell (in the case of a call) or buy (in the case of a put) the shares to (or from) the buyer of the option at the specified price upon the buyer's request.

It is very important to remember that an option gives the buyer only the faculty, but not the obligation to exercise the right to buy and sell the underlying that will be exercised only if convenient. As we'll see more clearly below, some kinds of options can be exercised even before the expiration, while others only on the expiration day. In any case, any kind of option can be resold before the expiration day to cash a profit or stop losses.

Now we can start to study the main features of an option:

- Underlying
- Premium
- Expiration
- Nature (call or put)
- Strike price
- Typology (European or American)

Underlying

The underlying is the stock, future, commodity, or Forex pair, or whatever other security, on which the option is exercised. The value of the option is linked directly to the underlying price. Do you remember the tulip options in Holland? In that case the underlying was tulips. So we have options not only on stocks and Forex, but on nearly whatever is listed in exchanges around the world. We also have options on stock indexes, so you can buy DJ Industrial Average, S&P 500, and NASDAQ. In this way, instead of studying the charts of single stocks and trading them separately, you can easily study only the chart of an index, and buy and sell futures and options on it. In the same way, if you have a diversified stock portfolio, you can buy puts on the stock index if you fear a drop in the stock market, to protect the gain you have made so far.

Also, if you want to cover yourself from the risk of currency fluctuations, you can use currency options. You can also use currency futures and/or options to speculate on a certain currency.

Premium

It is the price that an option buyer pays to purchase an option and at the same time the option seller cashes in by selling it. In other words, when you buy an option, you have to pay the seller a certain price called the premium. It's like an insurance premium: Once you have paid it, you are insured against unexpected events.

Expiration

After buying an option, you have several choices: reselling the option before expiration; exercising it before expiration (if the typology of the option allows you to do so); or keeping the option until expiration, when it will be automatically converted into the underlying. Let's imagine that we buy an option on the mini–S&P 500 future for $23.50. Because the future is worth $50 for each point, to know how much the option costs, you have to multiply 23.50 × 50. The price of the option is $1,175.

Nature (Call or Put)

As stated earlier, the option is divided into two categories: calls and puts. We discuss in detail in a later section how to buy or sell calls and puts and the related implications.

Strike Price

The strike price is the price at which the parties will exchange the underlying (if the option will be exercised), at a certain date. A synonym for the strike price is "option base" or just "the base." Each option on stocks controls one hundred shares in the U.S. market. If we want to buy an August call option on MSFT, we can choose among different strike prices available. Let's assume the stock's price is $26. If we buy the $25 strike price, the option price is $2.00 and we pay $2.00 × 100 = $200 to control 100 shares. If after a few days MSFT goes up and our option is worth $2.30, we can resell the option. We can also choose to exercise the option and to request the delivery of 100 shares at $25. The price of the stock is higher, and so we can enjoy the difference between $25 and the higher price, reselling MSFT shares. Furthermore, we can simply ask for the delivery at $25 (the strike price), and we can hold onto the MSFT stocks until they reach a higher level where we can decide to sell them.

Typology (European and American)

American-type options can be exercised in any moment before their expiration, while European-type options can be exercised only at their expiration.

Please notice that the words "American" and "European" are not referencing in any way the location of the stock exchange where the underlying is listed. They refer only to a definition that the economist Paul Samuelson coined to distinguish between options that could be exercised before their expiration and those that could not. The majority of options that have an underlying option in futures, both in Europe and the United States, belong to the first category. Instead, options on stock indexes and stock indexes futures are mostly of the European type, and they cannot be exercised before the expiration day. On that day, they proceed to make a cash settlement. In the United States, the expiration for stock options is fixed for the Saturday immediately after the third Friday of the expiration month.

CALLS AND PUTS: TWO SIDES OF THE OPTION UNIVERSE

Let's begin with some basic rules of thumb.

You can buy call options and your maximum risk is limited to the money you spent to buy the premium. In the worst case, you cannot lose more than the premium you paid. Of course, if you are a trader, and not a sucker, you are not supposed to wait until you lose all of the money you invested before your option expires with no value. If you are a trader, you close your option position as the price of the underlying reaches the level where you set your stop-loss.

Instead, if you write a put—that is selling short the naked option—your risk is unlimited because if the underlying soars, you are exposed to unlimited loss. We can examine what happens when we buy an option. We'll later see what happens in the case of naked short selling.

Through buying a call you can open a long position on the underlying; through buying a put you can open a short position on the underlying and you gain when the underlying price drops.

If you think that a stock price, or another underlying, will rise, you can buy a call. If you think that the stock price, or another underlying, will fall, you can buy a put. We can buy a call or a put by calling our broker, or online by using an Internet platform provided by the broker himself.

What is the most you can lose by buying an option? Whoever buys a call or a put will take a risk limited exclusively to the premium paid. A call buyer pays the seller a premium to obtain the right to purchase, by the expiration date or before, the underlying at the option strike price. Whoever buys a call bets on the rise of the underlying price.

STRIKE PRICES WITH RESPECT TO THE UNDERLYING PRICE

In the previous pages we spoke about options' basic concepts. Now, we explain the difference between situations in which an option can be considered OTM, ATM, or ITM.

You have probably seen these acronyms before. You will discover further on that they are very important for understanding the price dynamics of an option and value strategies you can build with them.

So, you will learn that options can be called with respect to their strike price, "in the money" (ITM), "out of the money" (OTM), or "at the money" (ATM). We'll also speak about options deep ITM or deep OTM.

ITM Options

A call option is named ITM, or "in the money," when its underlying price is higher than its strike price, or base. Follow the next example and this will become clearer.

Let's take a mini–S&P 500 call with a strike price of 1,100. If the S&P 500 future price will be higher than 1,100, then this call will be ITM. A put option is instead ITM when its underlying has a value inferior to its strike price. Following the last example, let's take a mini–S&P 500 put with a strike price of 1,100. If the S&P 500 future price were 1,090, the put would be considered ITM. Therefore, if the market shows an inferior value with respect to the strike price, the put is ITM and this means you are making money if you bought it when it was OTM.

The options that have as an underlying the equity indexes usually have a cash settlement. Other options that have as their underlying stocks are usually exercised when becoming ITM.

Therefore, a call buyer will usually find it convenient to exercise the option when the market soars and the underlying stocks' price will be higher than the strike price because the buyer will take advantage of the chance to buy the stocks at an inferior price compared to market price. On the contrary, a put buyer will find it very convenient to exercise the stocks if the stocks' price will be inferior to the strike price.

ATM Options

We speak about an ATM option, or "at the money," when its underlying shows a price equal to the strike price itself.

In this case a mini–S&P 500 option with a 1,150 strike is ATM when the mini–S&P 500 future is worth 1,150. This is true both for a call and for a put.

Usually, from a technical point of view, if the market would rise only 1 point beyond the strike price, we should consider the call just ITM. Instead, in the usual practice, we consider a 1,150 call still ATM, even though the underlying is worth 1,154. Why? Because despite the fact that the option shows a little intrinsic value, we still call an option with a strike price very close to the underlying price ATM, even though, technically, it is already ITM.

OTM Options

A call option is considered out of the money, OTM, when its underlying shows an inferior price compared to the strike price. So, a mini–S&P 500 call with a 1,100 strike price is OTM if the mini–S&P 500 future is worth 1,050, or in any case less than 1,100.

Instead, in the case of a put option, we can say that it is OTM if the underlying price is higher than the option strike price. Therefore, a mini–S&P 500 put with 1,100 price is OTM if the current price of the mini–S&P 500 future is 1,150, 1080, or in any case higher than 1,100.

OTM options, as we'll see in more detail in the next section, have a unique value that market makers and traders define as "time value." The time value is the probability that on the expiration day the option will expire ITM. If on the expiration day, the option is OTM, whoever bought the option before loses all of the premium and the option seller keeps it for himself.

Premium, Time, and Expiration Date

The further away the expiration date is, the higher the premium, speaking of the same strike price. What does this mean? It means that time, which in option pricing is a very important component, is considered something you have to pay for. Beginners shouldn't ever buy options with a very low residual time, meaning with only a few days left until the option expiration. This is why they usually don't have the experience to forecast strong directional movement within a few days, and so they could lose all the money invested in the option premium. This is a problem that commonly occurs if you buy OTM options with few days remaining until the expiration. An options seller is also called a "time seller," because he hopes that the option will expire OTM and waits for time to pass in his favor. The option buyer, instead, is fundamentally a time buyer because the options premium she paid incorporates the probability that as the time goes by her option will become ITM and generate a profit.

THE COMPONENTS OF AN OPTION PREMIUM

When dealing with the price of an option, the value we face forces us to make different considerations that are necessary to make efficient decisions in trading. Usually buying a stock does not present complex problems compared to an option because there are no temporal deadlines within that stock's lifetime. You just have to decide if you think its price is going to rise and buy it. Or, if it falls, sell it. The option, instead, has a life limited to its expiration day. In theory, you could keep a stock in your portfolio for years, and in case of loss, just hope that the price will rise again, giving you the chance to regain your losses (a strategy that I don't advise). The option, however, when it expires, ceases to exist, and you either earned money or you lost it.

The price of an option is determined by a series of variables on which the passing of time and the approach of the expiration date exert dynamic influence. Not taking these dynamics into consideration could cause you to lose money. Instead, by knowing them you can make a profit.

Those who buy options without considering these elements are not aware that at a certain point you risk losing all of the capital you invested to buy options. Since the expiration date of the options approaches and it is too late to return to the base, the chance that the options expire out of the money (OTM) is high.

Many people believe that it's very easy making money buying a call for a few dollars betting that it will produce outstanding profits because of the leverage implied in the option instruments. Instead, because of time decay, a concept we will explain later, many options expire out of the money. Those who sell naked options earn money and gain time value, with great satisfaction. The seller of the option takes a risk that exposes him, in theory, to unlimited losses. And so, the price of the premium is considered the right compensation for that risk.

What is the formula used to calculate this risk, and in general, what is the formula to calculate a fair price to value the price of an option? A few decades ago there were no clear answers to these questions and no formula existed, even though options have been bought and sold in America for the past 80 years. As we introduced when we spoke about the history of the option at the beginning of the chapter, until the 1970s there was confusion in the options market, even though these tools were widely available.

Also, U.S. financial institutions that sold options on stocks did not know exactly what value to give them, and so gave them a standardized price. Those days were when the first smart arbitrageurs could exploit the price windows of arbitrage and gain higher and secure returns. The income was huge and the risk was close to zero because they sold options covering their position of the underlying stock and took advantage of the misalignment of the market. Nowadays those kinds of wide arbitrage windows have disappeared due to the availability of sophisticated mathematical models and software that can evaluate differences between the options and the underlying, and these models work in real time with fast computers and technology that were not available thirty years ago.

It all started with Black and Scholes's studies of mathematical models, which are capable of giving more realistic prices for options. The models began working with known parameters, such as the remaining life of the option until its expiration, the interest rate on the market, the price of the underlying, the base or strike price of the call (or the base of the put), eventual dividends, and, above all, volatility.

Implied and Historical Volatility

Historical volatility is the measure of the fluctuation of the price around a central value. So the volatility gives us the magnitude of the movements of the underlying of an option in a determined period of time. The trader who buys options is in search of markets that have high volatility. The higher the volatility, the more you have the chance to earn higher returns, buying far strikes by investing a small amount.

Historical volatility is the one that comes from the applications of statistical formulae, studying the series of historical prices, which you can read on a chart.

Implied volatility is the one that the market maker attributes to the current market to request adequate compensation, because he acts as a counterpart to the other market participants. If the historical volatility of a certain market is, for example, 22, the market maker could use an implied volatility at 33 if imminent facts could affect the title. For example: they are awaiting the decision of an authority concerning a certain operation on the market. Whatever the end, the title that used to have a modest historical volatility now, for the market maker, is exposed to large oscillations in price because of the effect of the judgment of the authority. This is why implied volatility rises and the price of the options rises as a consequence.

Knowing these parameters and the price that the options have at that moment you can indirectly find the implied volatility.

Some people state that the Black and Scholes formula is outdated and does not fit the real price of an option. This is why many other models have been proposed, such as Cox, Ross, and Rubenstein and others that come from physics models. Also, Black and Scholes models stem from a model of statistical analysis used in physics to measure the expansion of gas particles.

Despite some scholars' criticism of the Black and Scholes model as imperfect, we believe it can still be considered a valid point of reference for the beginner to get a rough idea of the fair value of an option. In this way you can understand the most important dynamics that explain the option premium pricing and its two components known as intrinsic value and time value, which we'll study in detail below. In fact, Mr. Black himself argued, in an article published by *The Financial Analyst Journal* (August 1975), that the Black and Scholes model tended to underestimate the deep options

OTM and overestimate the deep ITM. Another small defect of the model is that it overestimates options with less than 90 days until expiration.

In any case, if you would like to become an expert in mathematical analysis and dynamic models for the pricing of an option, go right ahead. There are plenty of open debates in the scientific world and there is a Nobel Prize waiting for you if you figure it all out. But this book is not the right place to dig into those arguments because in this chapter I give you only an overview of the main features of options in case you are not very familiar with this subject.

To continue the path of our knowledge of the world of options, we have to study two basic concepts that need to be considered in every strategy you may use in the future: the value of time and the intrinsic value of an option. These are the two fundamental components in which you can divide the premium of an option.

INTRINSIC VALUE AND TIME VALUE

This represents the value of the option if you decide to exercise it right away. You can easily find it by comparing the base of the option with the market price of the underlying; it corresponds to the difference between the two values. That is why it exists only for ITM options, while OTM options do not have intrinsic value. Let's take an option for your favorite stock XY. For example, a call with base 5.00, of which the underlying (XY stock) at that moment is valued at 5.50: this means that if you are going to exercise the option now, we could buy the stock at 5.00 earning 0.50 by selling it at the same time. The intrinsic value corresponds to 0.50.

For a put, it works the same way. If we take a put on Microsoft (MSFT) with base 50 and in that moment the underlying stock is worth 45, then the intrinsic value of the option is 5 because the put is ITM and the difference between the base and the underlying is 5. Instead, if MSFT's price is 55 and we take a $50 strike price put, the option would be OTM and the intrinsic value would be zero because OTM options do not have intrinsic value, but instead, as we'll better understand later, they have only time value.

INTRINSIC VALUE

Remember that intrinsic values only exist for ITM options and that they correspond to the difference between the underlying and the base of the option itself.

Let's go back to the above example. Let's suppose that a call on our favorite XY stock with base 5.00 is worth 0.65, and the underlying stock price is 5.50. The intrinsic value is 0.50. The difference between 0.65, the price of the option, and 0.50, corresponds to the time value, which in this case is 0.15. It is the time value we are paying to take advantage of a rise in the stock price until the expiration date of the call option, as we will discover below.

We all know the saying "time is money," and we know that this is true. Well, in our case this saying is even more appropriate to the reality of trading because in the stock, commodity, and currency markets, speaking of options, you buy and sell time as if it were a real commodity or any other material good (bicycles, cars, or refrigerators).

Buying and Selling Time

When you buy long a call or a put OTM you are just buying time; the time you need to win your bet on the underlying price movement. When you purchase an ATM or ITM long call or put you are still purchasing time.

On the contrary, if you sell a call or a put short you are fundamentally selling time, by betting that the option will expire OTM before the expiration date.

There are listed options that allow you to buy more or less time. It means you can purchase options with one week, one month, two months, three months, six months, or nine months residual life before they expire. There are also special long expiration options called "Leaps." That means that they have different probabilities that on the expiration date they will result as ITM. And so, you have different opportunities to earn money. Obviously, the more time you wish to have your option active, the more time you need to buy. The more time you want to buy the more you pay for the option.

The table below shows the different real time prices for an 1170 strike price call option on a mini–S&P 500 future on October 27, 2010, with an underlying price equal to 1168. It's easy to understand how the more time you buy, the more expensive the option becomes.

November 2010	December 2010	January 2011	February 2011	March 2011
23.50	35.00	52.25	59.00	64.00

As you can see, the longer the residual life of the option, the more the price of the option rises.

In this example, our 1170 call with expiration in December costs 35 compared to the November expiration date, which costs only 23.50

because the option seller requires a higher premium to compensate for the higher risk he takes. The chance that the underlying, by the third Friday of December, will be higher than 1170 is elevated compared to the chance that this underlying will reach the same price in November.

The more time we have, the more the market can move in our favor.

Some people would pay just a few dollars to buy a call that is going to expire within one week. That means there are only five trading days to expire ITM. But pay attention! You could be very lucky if the S&P 500 soared within the next four days because your call could double its value. But the option could also easily expire OTM and the value of the premium would be zero within four days. It is true that if you buy an OTM call you spend less, but it's also true that the risk to lose the entire option premium is very high.

Unless you're an expert trader capable of using the most powerful analysis tools to predict the rise that has to happen in the next 72 hours (this is not that easy, even for veterans), don't buy options that are close to their expiration. For a beginner choosing between an option that expires within a few days and a game of Russian roulette, there is not much of a difference.

Otherwise, if we decide to invest more to buy an option with a residual life of 35 days the chances to gain will be higher because you have 35 days before the expiration for your call to become ITM.

Coming back to our table above, the 1170 S&P 500 call, even if it shows different expiration dates and prices, the dates and prices have something in common: they are all OTM, which means that their premium is composed of only the time value, because the underlying price is 1168.

Instead, an 1170 November call, when the underlying price is 1173, would cost 25.00. This call is ITM and the time value is 22 and the intrinsic value is 3.

Let's now consider an 1175 S&P 500 put, when the S&P 500 is at 1172. The put option costs 24.00, and so the calculation is very easy. The option is ITM and so we have an intrinsic value that corresponds to the strike price value minus the underlying price, which is equal to 3.

The time value is instead 21, and it is equal to the price of the option (24.00) minus the intrinsic value (3).

Well, now the whole picture is clearer and we are able to define the time value.

TIME VALUE

The time value corresponds in the ITM option to the premium of the option minus the intrinsic value.

In OTM and ATM options, the time value corresponds to the whole premium.

In the example above, all of the OTM and ATM options do not have an intrinsic value, and so their premium is represented only by the time value. Only in the case of ITM options is the premium composed partly of the time value and the intrinsic value.

The time value could also be considered as the part of the premium that the option's seller believes is fair enough to compensate him for the risk he takes selling the option. This risk is related to the residual time in which the option could become ITM and generate a loss to the seller.

For deep ITM options, far from their strike, the time value tends toward zero and the premium corresponds only to the intrinsic value because the option with a delta near 1 behaves like the underlying. For example, a 1070 call on an S&P 500 strike is a deep ITM if the underlying is 1200.

The Time Decay Phenomenon

When we decide to buy an option long, we are actually competing against time. If the underlying doesn't move and instead goes sideways, the option you bought loses a little piece of its initial value each day that passes. At the beginning, it loses its value slowly, but as time goes by, the loss rate speeds up until the premium reaches zero and the option expires out of the money.

The premium erosion phenomenon caused by the passage of time is called "time decay," and it is the worst enemy of an options buyer. On the contrary, whoever sells naked options hopes the market remains stable or rises, in the case of naked put selling; or, remains stable or falls, in the case of naked call selling. To sell an option short is called "options writing."

The option short seller hopes to sell the option at a certain price and, because of the time decay effect, to repurchase it at a lower price, or keep the entire premium if the option expires out of the money.

In Figure 13.1 we can see a five-step binomial tree of an 1170 mini–S&P 500 November call and put option with a residual time of 23 days till expiration. It was bought at 24.50 (24.69 is the value in the theoretical binomial tree model) when the underlying, the mini–S&P 500 future, was at 1170.

The amount you spent to buy one 1170 mini–S&P 500 November call option is easily obtained multiplying the option price × $50. In our case it is 24.50 × $50 = $1225. $50 is always the multiplier for price of a mini–S&P 500 call or put option.

The binomial tree is an interesting tool to evaluate the course of the variation of the price of an option, looking both to the variation of the underlying and the time decay, assuming that volatility will not change. Even though the method of the binomial tree is based on a different formula compared to the one used by Black & Scholes, it can be, in any case, used to understand what happens to the price of an option step by step in case of the underlying price fluctuations, or in case of no variation of the underlying price.

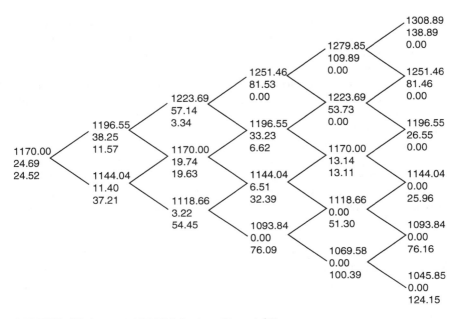

FIGURE 13.1 Mini–S&P 500 Options Binomial Tree

We will follow the course of the call in some of the steps of the tree and then comment on the implications.

We find on the left the prices of the underlying from which we will begin our simulation, 1170, and immediately below on the right the value of the same call if the underlying after 4.6 days reaches 1144. In this case, the option price would be 11.40. Below 11.40 you will find that 37.21 is the price of an 1170 put after 4.6 days if the underlying reaches 1144, bought when the underlying was 1170. For the rest of the tree the sequences stay the same: first the value of the underlying, directly after that the value of the 1170 call, and below, the value of the 1170 put. Obviously, the strike price is always the same in the above tree, while the value of the underlying and the option's price change in each step of the tree.

After the initial price, we can follow the evolution of the call or of a put with the same strike price every 4.6 days. In fact, the next five steps of the tree ascending or descending from the initial price of 1170 tells us how the value of the option changes because of the passage of time, both in the case of the underlying price rising or falling.

For example, if the mini–S&P 500 future, that is the underlying, after the first step of 4.6 days, goes down to 1144, then the call price goes from 24.69 to 11.40. But if after another 4.6 days the S&P500 goes back to 1170, then the call is valued only 19.74 against its initial value of 24.69, when the underlying was worth the same price. If the underlying goes back to the same price at which we bought the option, then the call's premium value is less. It is not 24.69 anymore, but rather 19.74, according to our Cox, Ross, and Rubenstein tree model.

It means that we have lost about 20 percent of our initial investment.

This situation is of course positive for those who sell us the option and wrote the 1170 call. If the seller would close the operation he would close with a good profit.

Obviously if the implicit volatility changes after one or two steps, the value of the option could be altered. If the volatility rises, the values of the options rise, if instead it falls, then the call and the put price will fall as well. The Vega, as we'll see better below, helps us in understanding how much change in the price of an option is subject to the change of the implied volatility.

This is the binomial tree that gives us the chance to monitor the options price fluctuations of the put every 4.5 days, if we compute it with five steps. Naturally you could use the same software that creates the binomial tree to trace many more steps to monitor the changes of the price more frequently, not only concerning the time, but also the smallest incremental differences of price.

They usually consider the Cox, Ross, and Rubenstein binomial tree formula more precise to calculate an American option, such as options having a stock as the underlying that can be exercised in any moment before expiration. Instead the example we are dealing with is about an equity index option, which is considered a European option, even if the index is the American S&P 500. Nonetheless, the binomial tree is very useful to show to a beginner the option's price fluctuations and the time decay, theta phenomenon.

If the underlying's price moves in the direction you expect (rising if you bought a call and falling if you bought a put) then the value of the option premium goes up. But also in this case you have to take under control the time decay, because if there are only nine days left to the expiration date and you bought before an 1170 mini–S&P 500 November call at 24.50, and now the underlying is at 1196, the call is worth 33.23. Now you're making a lot of money because the option premium earned about 35 percent. The initial option price was 24.69 and it is now worth 33.23. So if you resell it you can cash a profit of 35 percent. There are only nine days left before the option expires and if the market stops rising and remains around 1196, the call devalues very quickly and its value, approaching the expiration date, will be near 26, which is the intrinsic value.

But because we bought the option at 24.50, the profit in this case would only be 1.50 instead of 8.54 per option. The final dollar value can be obtained by multiplying each option price by $50.

So if you sold your option before nine days to the expiration you would earn 8.54 × $50 = $427 per call purchased.

Instead, if you wait another four days and the price of the mini–S&P 500 comes back to 1170, and you do not resell your call, you would earn only $1.50 × $50 = $75 for each option you bought.

As you can see in this case, time decay is fatal for a trader who doesn't sell the option on time. The measure of an option's depreciation, caused by the passage of time, is called theta. It is one of the most important parameters of option pricing, known under the collective name of Greeks, and it allows us to evaluate the premium price valuation in an analytical way. The Greeks are worth studying.

THE GREEKS: FUNDAMENTAL PARAMETERS OF OPTIONS

How can we evaluate the price of an option? Pricing an option, which is finding the closest price to the theoretical fair value, is a very important matter. It is fundamental to create effective strategies to consistently gain in financial markets. As we saw in the previous section, when the option's price moves away from its theoretical value, the operators and the arbitrageurs immediately step into the market to take advantage of the "price window" that was created. Today these arbitrage windows are very small thanks to the diffusion of sophisticated mathematical models that operate in real time on computers.

We're now ready to examine the matter of evaluating the price of an option. Remember, to determine the profile of risk, or the profile of profit of an option, including its value after changes (increasing or decreasing) of the underlying, we have to consider the following variables:

1. The option's residual life (time to expiration)
2. Volatility
3. Interest rate
4. Dividends

These variables directly determine the price of options and, of course, the price of the options changes according to these variables. We use four parameters, known as the Greeks.

1. Delta
2. Gamma

3. Theta

4. Vega

Delta and gamma give us the size of the value of our expositions compared to the alterations of the underlying.

Theta gives us the variation of the premium compared to the timeline, allowing us to have a clear image of the *time decay* phenomenon.

Vega gives us the size of the value of our positions compared to the change of implicit volatility.

Each of these parameters helps us to accurately measure the risk profile and the reactivity of an option, and measure its price fluctuations after any of the parameters change.

Correctly using the Greeks allows professional traders to create precise management strategies for a complex position based on options, and also allows the trader to strategically manage and adjust the risk profile.

These parameters give us a detailed image of gains and losses that our position could generate subject to a change in the value of the underlying and the implied volatility.

Delta

Delta is a very important parameter that both professional and beginner options traders must pay attention to.

D E L T A

Delta is the ratio between the variation of an option's price value and the variations of the underlying's price.

Delta helps us to understand how much the option's price changes compared to the movements of the underlying's price. It has a value between zero and one, or if we multiply the result by a hundred, between one and a hundred. If an option has a delta of 0.5 or 50 it means that when its underlying varies by 100 cents the option will change by half, in this case, 50 cents.

For instance, if a call on an IBM stock has 0.5 delta it means that if the IBM stock rises five points, the option revalues itself by 2.5 points. Or, consider a put on the mini–S&P 500 future with a delta equal to 50. If the future decreases 10 points, the option will revalue itself by five points. Every ATM option with a strike price very close to the value of the underlying has a delta equal to 0.5 or 50.

Stocks and futures by definition, regardless of the underlying, always have a delta equal to 100. This is because they are the underlying of an option and the point of reference for the option price change. So, by definition, they have a fixed delta.

Options, instead, have a dynamic delta. For example, let's consider buying an 1150 strike mini–S&P 500 call, when the mini–S&P 500 future is at 1150. The call is ATM, and so the delta is equal to 0.5; but if the market rises by 40 full points and reaches 1190, the delta of the option rises in a more than proportional way. As the call becomes more and more ITM, its delta will always increase. After a big increase of mini–S&P 500 futures, we notice that the variations of the option's price are almost the same as the variations of the underlying's price. This means that the option, by now, will have a delta equal to 1.

So the deep ITM options have a delta equal to 1. To summarize:

- Stocks, futures, and stock indexes have a fixed delta equal to 100.
- Options instead have a variable, dynamic delta.

Delta can show a positive or negative value, according to Table 13.1.

For example: if your favorite stock XY's price is currently 25.00, a call with base 25.00 will have a delta of +50 or +0.5; a put with base 25.00 will have a delta of –50 or –0.5. This distinction is useful for calculating the total delta of a position made by options with different bases, which we can decide to hedge, or balance with a position that has the same delta but with the opposite sign.

What is the benefit of buying a call with total delta +100 and then selling an S&P 500 with a delta equal to −100? It is very important to carry out delta *neutral hedging* and create strategies such as the synthetic straddle.

Variations of Delta Compared to Time The passage of time has a specific impact on the value of an option's delta. The delta of OTM options tend toward zero as time goes on, so an option's OTM reacts less to the changes of the underlying as we approach the expiration date. The more time we have until the expiration date, the more the delta will have the chance to increase. The delta of an ITM option will increase more and

TABLE 13.1 Positive and Negative Values of the Delta

Underlying Delta	Underlying Delta	Underlying Delta
Long stock or future +	Long call +	Long put −
Short stock or future −	Short call −	Short put +

more tending to 1 as time goes by. The bigger the time value of an ITM option is, the lower its delta.

Buying Strategies Based on Delta   Let's assume our option to be on a price support level of stock XY, a very liquid share. The stock chart shows a double bottom, on a very strong QPL, at 19.05.

We think that this is a very good short-term entry signal, a good occasion for making fast profits. At this very moment, the stock's price is $19. We want to buy a call, thinking that the stock will rise within the next few weeks.

If we wonder which is the best strike price to buy, let's look at the different options shown in Table 13.2. Imagine it's December.

A beginner would probably buy the January 20 strike price or the March 22.5 strike price, because they are the less expensive options.

A professional trader would probably buy a January 15 call at 4.00, which doesn't have a time value, or a January 17.5 call at 2, because it has a time value equal to 0.50.

Why? Because the delta of these two options is very high and so it will react strongly to a change of the underlying.

Instead, a beginner would be tempted to buy an option that costs less, such as the 22.5 strike in March. The lower the delta, the more we need an important movement to increase our call value because delta represents the probability that at the expiration date the option will be ITM.

Buying an option with a high delta gives us the chance to make more money on a certain movement of the underlying compared to an option with an inferior delta.

Hedging Ratio and Delta Neutral   As we saw, we can build a delta position considering the positive or negative sign of delta. Let's assume we have bought two mini–S&P 500 calls with a delta equal to 0.5: one mini–S&P 500 call with a delta equal to 0.3 and another call on the same index future with a delta equal to 0.7.

TABLE 13.2   Strikes and Prices

January Options on Stock XY	Price	March Options on Stock XY	Price
Base 15	4.0	Base 17.5	2.50
Base 17.5	2.0	Base 20	1 and 1/8
Base 20	0.5	Base 22.5	0.50

In this case, the total delta of the calls is +2.0. We have four long calls with a different strike price, but they all have a delta with a positive sign. To hedge our mini–S&P 500 calls position and the underlying, we have to proceed as follows: first, we have to remember that one mini–S&P 500 future contract always has a fixed delta equal to one, just because it is a future, as we have already explained above.

Second, we have to calculate the total delta of our option position, which we have already seen is +2.00. The positive delta of our option position will be hedged, selling short two mini–S&P 500 future contracts, which have a total delta value equal to −2.00.

Using a good broker you usually don't pay double margins because the broker will apply a cross margin. This means that the broker will set aside the margin from your account only once and not twice.

If you want to be consistent with your hedging program, you should buy options with the same expiration date of mini–S&P 500 futures, meaning only on quarterly closures on the third Friday of December, March, June, and September.

You may want to carry out this hedging activity when you wish to lock in the profit you have made both on your option position and when for some reason you do not want to liquidate it before the option expiration date.

So, according to what we have learned so far, a portfolio position composed of stocks, futures, commodities, and currencies can be hedged by an option position having the same net delta value with an opposite sign. We usually say that a future position is equal to a determinate net delta position.

If we're going to calculate the net exposure of our option's position, we need to calculate the number of futures contracts to buy or sell to create a so-called *delta neutral position*. A delta neutral position is based on the equivalence between the option's delta and the future's delta. But they have to have opposite signs. Please refer to Table 10.2.

The *hedge ratio* tells us how many futures contracts or options we should buy or sell to have a hedged portfolio.

Delta Neutral Trading

The goal of setting up a delta neutral structure is to build a position in which we don't lose money within certain rules, whether the market rises or falls.

In fact, at the beginning what we lose with one position will be gained by the other side of our strategy.

After the underlying price breaks, in whatever direction, the lower or upper delta neutral strategy, then we start making money because one side of the strategy increases its total delta, causing the trader to profit.

A neutral delta position can generate a profit under specific conditions:

- When the break-even point of the structure is broken within a superior or inferior range
- When the neutral delta's structure is used to create a dynamic hedging to lock in the profits on a position that was previously gaining, avoiding the eventual fall, either overnight or during open market sessions, which could erode the profit we made

The hedge ratio is very well known amongst professional options traders and it is largely used by portfolio managers.

Gamma

This is a much more technical parameter, used above all by fund managers and by professional traders. It is important to understand gamma to have a greater comprehension of options.

Gamma measures the variation speed of delta. It measures the changing rate of delta related to an underlying variation. It shows precisely how delta can vary, either slower or faster, compared to the underlying.

GAMMA

The option gamma measures the change in an option's delta for a one-point change in the underlying price.

The option gamma of both long calls and puts is always positive. The option gamma of both short calls and puts is always negative.

Having positive gamma means that the delta rises as the underlying price increases. On the contrary the delta will decrease as the underlying price falls.

A positive gamma means that the option's delta will rise, as well as the underlying's price, or vice versa, when the delta will fall following a decrease of the underlying's price.

The option's delta with negative gamma will decrease while the underlying increases, and will rise as the underlying falls.

Long puts and calls have a positive correlation with the underlying's movements, which is what we say when they both have a positive gamma.

In fact, a long call grows in price at the increase of prices, and a long put behaves the same when the underlying's price goes down. The short call and short put positions have a negative gamma because they show negative correlations with the market's movements.

The gamma can be expressed in U.S. dollars, for it is a measure of the variations of the delta of an option for every change of a dollar in the price of the underlying stock.

It can also be represented as a number between zero and one. If an option's delta is equal to 0.48 and its gamma is 0.07, a point of increment of the underlying stock will lead, in theory, to an increment of 0.07 of the delta, equal to $0.48 + 0.07 = 0.55$.

As we have already explained, stocks and futures have a fixed delta that doesn't change with the variations of the underlying's price, so they have no gamma.

Since the gamma gives the measure of the speed of the changing of an option's delta, on a mathematical level we can consider it as the second derivative of the option's price related to its underlying price. This, in physics, corresponds more or less to the concept of acceleration of the option's price retagged to the underlying. Therefore, the higher the gamma is, the faster the speed of change for the delta.

The gamma varies at the two extremities of the price's band of an option, which has on one side an option deep in the money, and on the other side an option far out of the money. *Deep in the money* means that the option strike is very far from its underlying's price, having already been ITM for a while. *Deep out of the money* is an option with an OTM strike price that is very far from its underlying's price. For these two kinds of options, the gamma is very close to zero because there is a low reactivity of the delta compared to the underlying's price variation.

Let's consider a 1030 mini–S&P 500 call option. If the mini–S&P 500 future price is 1300, the change of one point of the underlying will not have any influence on the changing of the option's delta, which is already close to 100 because the option is deep in the money.

In the same way, for a mini–S&P 500 call 1200, if the mini–S&P 500 future price is 900, the gamma will be equal to zero because the delta is close to zero and the changing of an underlying's unit doesn't have any effect on the increment of the option's delta.

ATM options have the largest gamma. As time goes by, the gamma of ATM options increases, while the gamma of deep ITM and OTM options decreases.

Regarding the relationship between gamma and volatility, when volatility decreases ATM option's, gamma rises, and the gamma of deep ITM and OTM options fall.

Theta

Theta is a fundamental parameter, both for the professional trader and the beginner who are approaching the world of options. It measures the option's premium's loss of value for each day that passes approaching the expiration date.

Theta is very similar to our daily life because the stock exchange can be a metaphor for the game of life. However, in the financial markets everything happens more quickly: one can make huge gains or terrible losses in a very short amount of time. But if you consider the hectic rhythm of everyday life you can easily find many similarities.

Each day, people lose opportunities. Often times it's because people do not have effective strategies and do not achieve success because they procrastinate taking action, thinking they have unlimited time at their disposal, as if they believe themselves immortal.

Carlos Castaneda alludes to this problem, observing that the contemporary human being hardly takes responsibility for his own actions, and is affected by the immortality syndrome. Man, usually, prefers to wait for tomorrow instead of making a crucial decision now that could change his life forever.

This also happens in the options world. If today we're gaining on our long call position because the market rises, and we don't evaluate the risk's profile of keeping this position open, being exposed to the risk of time decay, we could easily see an initial significant gain quickly mutate into a loss. Not only could we lose what we earned, but also the entire premium we paid.

It can happen that we buy an OTM option while the market is moving strongly in the direction we predicted, but since the expiration date is near we decide to not immediately close the position, and we keep it hoping that the market will rise even more.

Let's suppose it's the Friday afternoon one week prior to the technical expiration. We made a profit with some calls still slightly OTM. At this point it is very important to immediately sell them to lock in the profits.

If we're led by greed and we wait for the market to rise even more, we could, if we're lucky, make an extra gain. But the risk of losing our entire paid premium is high.

In fact, the call is still OTM. Even if it were to give us a profit, it could expire OTM in seven days, an eventuality that frequently happens. So the theta supplies us with a very precise measure to calculate how much the premium will deplete each day until it reaches zero at the expiration day.

As we said before, this phenomenon is called *time decay*, and it is the worst enemy of the option's buyer.

The option's seller, instead, is favored by time decay, because he hopes for the option to expire OTM. So whoever writes a naked call, which is selling short a call, or a naked put, which is selling short a put, should study the time decay process. This is very important to know how much the premium he cashed by selling a call or put will deplete, and also to know what his profit is from the passage of time in case he decides to buy the option before the expiration date, closing the short position.

The phenomenon of time decay is not 1:1 proportional to the passage of time until the expiration. Rather it is more than proportional as we approach the expiration date.

Without delving into the mathematical formula that regulates the entire system, so as not to bore the beginners, we can roughly state that the rate of decrease of the option's premium is proportional to the square root of the remaining time before the expiration date.

In any case, we can say that the premium of an option with a sufficient temporal length, corresponding to about three months, will lose around a third of its value in the first half of its residual life, while two thirds will be lost in the second half.

Usually an ATM option has a more elevated theta and so, compared to the OTM options, loses more value as time passes.

At this point, it is very important to talk about the implications that come from trading positions based on the options compared to those based on the fixed delta. In other words, is it better directly trading the underlying, which can be a future, a stock, or a currency, or trading the options?

Traders that buy or sell stocks, futures, or currencies have, of course, the advantage of not being victims of time decay, and so even if the market is congested and the volatility decreases for a short or long period, the value of their position will at least not be influenced by the passage of time, because a stock doesn't have an expiration date, as options do.

In the United States all the brokers grant, at the very least, the traditional leverage of 2. This means that they are allowed to buy stocks after a previous margin's payment of 50 percent, allowing them to buy (if you have $50,000) stocks for a value of $100,000. Profits or losses in the trading will have to be multiplied for 2, so they will have a leverage of 2.

The advantage that an options trader gets from the choice to trade options instead of the underlying is that he can use a higher leverage. This allows the trader to amplify the profits. But the trader always has to pay attention to risk management, considering the maximum loss he would be willing to take.

The big disadvantage for an option buyer is represented by the time decay phenomena, which reveals itself to be very dangerous for beginners who tend to undervalue it. They expose themselves to higher risks of loss when the option they bought expires OTM, or even after changing into ITM, they see the price pull back and the option becomes OTM once again.

This is why we suggest that beginning traders do not ever buy options that have less than 30 days until expiration. This is because, even though the strategy could be impeccable, a change in volatility, or market congestion could put the trader in a critical situation and cause substantial losses.

This can happen, not because of a mistake in the chart analysis, but rather from the time decay process and the approach of the option's expiration date.

If we're buying an option with a residual life of only 10 days, the following becomes very probable: The entrance strategy is correct, but the market stops moving and goes sideways. Then it begins moving again a few days before expiration. The target is not achieved and the option expires OTM.

Instead, if we have many days before expiration, let's suppose at least 30 to 45, then even if the market stops moving, the temporary loss of the option's value, due to the time decay, will not weigh upon the final result of the operation that much.

Short Strangle and Improper Use of the Theta When Selling Options Now we introduce a very popular options strategy called "short strangle" and we explain how theta dramatically impacts this strategy. Unfortunately this strategy is sometimes carried out without the complete understanding of all the implications of the strategy itself. So we'll review the main features and any problems connected to it. We advise that you first familiarize yourself with long strangle strategies, and only after consider whether to move to short strangle.

Concerning the short strangle strategy, several manuals or option classes state that a good and safe strategy to consistently make money is to sell options with a strike price far from the underlying, because the time decay phenomenon, or theta, allows the seller of a naked option to expect a considerable monthly profit.

This strategy is based on simultaneously selling one relatively deep OTM naked call and one relatively deep OTM naked put. So selling very far bases from that actual underlying price, the followers of this strategy count to take the entire premium they received from the options buyer for themselves, betting that the two options will expire OTM.

This could be true, but be aware that volatility sometimes dramatically changes unexpectedly. Remember that if you have specific analysis tools that can indicate the historical locality (historical volatility), you can build a proper index that will allow you to have an *outlook* about a probable prevision of the price's future movements.

An unexpected sharp rise in volatility is always possible. In this case short strangle can create many problems if you do not have a good hedging strategy.

The problem arises when, after an important directional movement, the price of the underlying breaks the strike price of the option you wrote, and so expires ITM, causing a loss to your position.

In this case, there's no limit to the losses that the inexperienced trader could face if he cannot manage the mechanism of hedging.

Let's suppose we sell ten 1220 mini–S&P 500 calls when the mini–S&P 500 price is 1150, and ten 1080 mini–S&P 500 puts opening a short strangle with 70 points of distance from the underlying's price. Let's assume that the naked options you sold will expire within a month, and that the market, after seven days, begins a very strong directional movement, approaching 1215 by moving up 65 points in five trading days.

Now we have two basic possibilities:

The first is to hope that the mini–S&P 500 future price will fall, so the option will expire OTM.

The second is to hedge the position by opening the opposite sign's delta position, such as buying some mini–S&P 500 futures to balance the total delta of the naked options sold, only for the side that is causing a loss.

Regarding the first hypothesis, some people think, "Let's wait for a market correction because, since it hit 1220, it will sooner or later drop back down." Because of the time decay, you might see the sold 1220 mini–S&P 500 calls expiring OTM.

Also, if the market pulls back, returning to the original price level in which you opened the short strangle, you might decide to close the strangle at the same price in which you opened it, with a small loss or profit, depending on the implied volatility change. In trading, it is very important to have both an "A" plan and a "B" plan, thinking not just about the best possible outcome, but also creating a strategy that protects against huge losses.

When the underlying price is near the strike price of the 1220 mini–S&P 500 call you sold, you should cover it by buying one mini–S&P 500 future for each call sold. By doing so, you'll keep all the profits coming from the naked call when the option expires, even if the underlying price rises more. The resulting losses for the option position coming from the difference between the underlying price and the sold base would be entirely offset by the profit coming from the mini–S&P 500 futures profit.

This strategy helps us to make a profit even if the market increases to 1280. The market has to be strictly monitored to avoid losses that can occur if the underlying's price pulls back after buying mini–S&P 500 futures.

Some traders believe that short strangle, based on the time decay, is one of the easiest ways to make a profit. The hidden risks of this strategy were partly discussed. In any case, it is very important to use this strategy only after analyzing the market, and only after having developed models to forecast volatility changes to avoid an increase in the implied volatility when we sell naked options.

Making money by selling *naked options* is a great strategy, but to be successful you need the knowledge to choose the best strike prices to sell, where the price is more likely to find a strong resistance level. Quantum Trading can provide you with very good information about this subject, as we'll see in Chapter 14. To succeed in selling strangles, you need to know how to manage the embedded risk and the strategy to cover short positions in case the price breaks the options' strike price.

Vega

Vega is the measure of how much the variation of a unit in implied volatility will affect the option's price variation. For U.S. stock options it is usually measured in cents. If an option has a vega of $0.25, it means that the option's price will change $0.25 for every full percentage point change in volatility.

We've already talked about implied volatility, and vega is a parameter connected to implied volatility. Therefore, the options trader needs to be aware that a change in implied volatility is fundamental for a variation in the premium value of the sold or bought options. The implied volatility and its variations are to be kept under careful observation.

In fact, a rise in volatility leads to an increment in a premium's options. On the contrary, a fall in volatility will lead to a drop in the option's price.

The option seller hopes for a drop in implied volatility because it could allow him to immediately close the position and profit even though the underlying's price doesn't move very much.

Let's suppose that the volatility in mini–S&P 500 call options is equal to 22 and the market, after a few days, increases its volatility by hitting 30, and then the price returns to the same level where it was when two traders reciprocally bought and sold a certain option. This leads to the following situation for the buyer and the seller.

The seller is losing money because the price of the option increased, and so if he would like to buy back the option he previously sold, he would have to pay a higher price.

The options buyer is in the opposite situation, because he's making money. In fact, if the trader would like to resell the options he bought before, he would profit from the rising value of the option's price, resulting from an increase in implied volatility.

Vega, then, provides the measurement of how much the option's premium will rise or fall with respect to the variation of the implied volatility. Usually vega augments when we approach the option's expiration date. This is because a rise in implicit volatility has a higher impact on an option with a shorter life.

In trader language, we say that a short call or short put position has a negative vega because it takes profit from a decrease in the implicit volatility. Instead, a long call or long put position has a positive vega because we gain, upfront, from an increase in implicit volatility.

Most beginner traders believe that on short call or short put positions it is only possible to gain, thanks to the effect of the passage of time.

The closer we get to the expiration date, the faster the option's premium approaches zero, and the option expires OTM, due to the effect of the theta.

The vega, on the contrary, has a fundamental effect on the curve of profit or loss of the naked selling of options because a sharp diminishment of implied volatility can generate very large profits.

Thus, if both vega and implied volatility fall sharply, then a short call position can be closed for a lower price than the selling price. Then the position can be closed with good profits.

The main features of Call and Put options can be found in Table 13.3.

TABLE 13.3 Call and Put Options

	Call	**Put**
Buyer (Long)	Right to buy stocks at strike price	Right to sell options at strike price
	Obligation to pay the premium	Obligation to pay the premium
	Profit from an increase of underlying price	Profit from a decrease in underlying price
	Potential gains: Unlimited	Potential gains: Unlimited
	Potential losses: Limited	Potential losses: Limited
Seller (Short)	Duty to sell stocks at strike price, if the option is exercised	Duty to buy stocks at the strike price, if the option is exercised
	Right to cash in the option's premium	Right to cash in the option's premium
	Profits if the underlying price remains stable or decreases	Profits if the underlying price remains stable or increases
	Potential gains: Limited to the premium	Potential gains: Limited to the premium
	Potential losses: Unlimited	Potential losses: Unlimited

CHAPTER 14

Options Strategies with Quantum Trading Tools

W
hile reviewing Chapter 13, we begin to appreciate the beauty and the power of option strategies based on Quantum Trading techniques. If you are primarily an option buyer, or you prefer to be an option seller, you can appreciate the help provided by QPLs and QPLSHs to correctly choose the best strike price to build up an effective options strategy to trade your favorite shares, stock indexes, commodities, or currencies. In addition, you can learn about using TAs and searching for a Quantum Parousia to find the best time for opening a profitable option strategy.

To be a very profitable option trader, it's very useful to have an understanding of the price deflection phenomenon stemming from Einstein's light-deflection theory and the quantum leap idea that are behind our Quantum Trading model, which conceives of the price as if it were an electron, jumping from one energy level orbital to another.

QPLSHs: A POWERFUL INDICATOR TO BUY AND SELL OPTIONS

As we have studied in the previous chapters, quantum price lines show significant support and resistance levels. We can use QPLs and QPLSHs as very precise indicators for choosing the strike price to be sold in many strategies such as, but not limited to, a short strangle strategy. We will sell the strikes lying on a significant support and resistance that is difficult to break.

Furthermore, QPLs can indirectly tell us something about volatility. If a stock price has been dropping for a long time, but is approaching a strong QPL, entelechy, or Quantum Parousia, then that should be a very strong support for the price. Usually the postcrash price of the call is much cheaper than the put price, because put implied volatility is higher than call implied volatility. So, it could be smart to buy a call option on the price level suggested by Quantum Trading tools. If you buy a call option at this moment, forecasting a reversal, the implied volatility for a call should be very low compared to the value that it could reach within the next few days or weeks if a rally has occurred. You can take advantage of two elements: a rise in the underlying price and an increase in volatility. The combination of these two different elements would be explosive and this could make your long call option price soar.

Let's take a look at the EUR-USD spot chart (see Figure 14.1) that is presented in previous chapters, as well. When the price is approaching point C on "P" QPL, we can monitor the call price of EUR-USD. When the price is near the QPL, passing point C at 1.2333, we'll follow the price dynamics. The more aggressive trader can wait for a Quantum Parousia at point C, buying the long call when the underlying price is near the QPL at point C. The more conservative traders can wait for two or three 60-minute bar charts higher highs and then buy a call ATM or slightly ITM. Both types of traders will take advantage of a strong reaction of the delta. In this way the timing for opening a long call strategy will be perfect, because you will take advantage of an explosion in the call's implied volatility, and the upcoming rally.

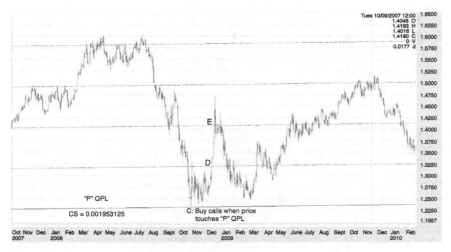

FIGURE 14.1   EUR-USD and Option Strategies Based on QPLs

An option buyer should always buy a call when volatility is very low and resell it when the implied volatility is high. Instead, beginners tend to do the opposite: they buy option when volatility is high and they resell it when it decreases.

As you surely remember, if you trade options, the theta process could be an issue. It's very important to have a clear idea of what price will close your long call strategy, because if the market already has risen and you bet that it will keep rising, you could lose the majority of the profits you have earned so far, as we explain in Chapter 13.

So you can use both QPLs and QPLSHs as an exit target to close the long call position as well as TAs. You just draw on the EUR-USD chart all the different quantum price lines, and they provide you with the best price level to resell your call. In this case, because we used "P" QPL as a support for opening a long strategy, and the next 45-degree "P" QPLSH is around 1.32 (Point D in Figure 14.1), we'll resell our call option exactly when the EUR-USD price reaches this level or, even better, slightly before.

But if we wait another few days we'll have earned much more, because the euro strongly rallied against the U.S. dollar, also reaching the next 45-degree "P" QPLSH around 1.4079 (Point E in Figure 14.1) only the day after. Wow! It was one of the biggest two-day rallies for the euro. You could have waited for this 90-degree "P" QPLSH higher target. It depends on how aggressive your trading style is.

Quantum Price Lines and Options

QPLs and QPLSHs provide us with amazing targets for buying and selling options. Many classic options strategies can be much more profitable using quantum price lines as starting points for opening a position or as a target price to close a strategy, thereby maximizing the return.

Volatility for both call and put can dramatically change when the underlying price approaches important QPLs and QPLSHs. TAs can enhance your ability to be precise in forecasting a change of the trend, and help you to buy an option when implied volatility is lower and it costs less.

Other nondirectional strategies such as, but not only, long straddle and long strangle, can be successfully used, increasing the return and lowering risk, considering the price level pointed out by Quantum Price Lines.

The same approach can be used to open a long put position. If you take a look at Figure 14.2, showing the S&P 500 futures chart that is also presented in Chapter 5, you can easily figure out that the all-time top, on October 11, 2007, was made just on an "S" QPL. On this level the put options were very cheap, and because this "S" QPL was very powerful, we were

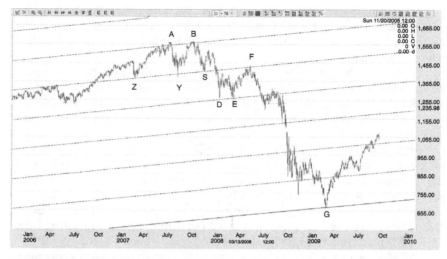

FIGURE 14.2 S&P 500 Future All-Time Historical Top with 40 QPLSHs (CS = 4)

expecting a reversal and a big drop in this index. This was the best level (point B) to open a long put position, buying options that expire after three to six months because of the low implied volatility and the very inexpensive price needed to open a long put strategy. The targets to close a relatively short-term operation are indicated by the QPLSHs below the "S" quantum price line. If you examine Figure 14.2, you will realize that you can close your position at point S around 1413. So when the underlying price reaches this level you can resell your put, enjoying a very nice profit.

If you prefer instead option short selling, using Quantum Trading tools you can develop many strategies based on naked option writing.

If you have never traded an option before, please do not begin your option trader career by writing an option, but instead please study all the strategies, try making trades on paper first, and then start with some long call and put positions, just to acquaint yourself with the options world.

You can start selling ATM or slightly ITM puts if you bet on a rising or a sideways trend of the underlying on a certain support level, which you can spot using QPLs. If a security price has been dropping for a long time, but is approaching a strong QPL, or better, an entelechy, then that should be a very strong support for the price; the put option should be quite expensive compared to the call. If you like short option writing, and you are not an absolute beginner, it is the best moment to sell naked put options, because the market maker considers acting as your counterpart for the short side of the market to be more risky than acting for the long side.

When the market rebounds, the implied volatility of the put will decrease. So it will be more convenient for the option writer to repurchase the put at a lower price. This is due in part to the rally that pushes the price of the underlying far from the option strike you have sold, and in part to the lower implied volatility.

In the same way, QPLs and QPLSHs can be used to choose the best level to sell the different wings of a short strangle.

VOLATILITY-BASED SPREADS

In the previous section we examined some option basics and we mentioned delta hedging and the synthetic straddle. Now we are ready to review some volatility spreads.

Volatility is a very important variable upon which several different strategies can be proposed. For instance, strategies based on increased volatility, such as long synthetic straddle and long straddle, tend to open concurrent positions with different options to gain when the market increases its volatility after a fast underlying variation. When the market moves nervously with widely varying prices, volatility increases and option prices are revaluated accordingly. Instead, a strategy like short strangle can make money after a decrease in implied volatility.

Long Straddle Strategy

Long straddle is a very useful strategy to use when we expect volatility to increase in the following days, but we are not sure if the market will go up or down. Therefore, if we attend to a crucial phase, a significant rise or fall of the price, we can take profits regardless of the market direction.

Building a Long Straddle Strategy

How can we build this strategy? We just buy an equal number of calls and puts at the same time with the same base, which usually, but not always, can be ATM. This strategy can be catalogued among the "nondirectional strategies," which take profit from a huge movement of the market, whatever its direction.

A long straddle breakeven point is calculated as follows:

Lower breakeven point = strike price − (call price + put price)
Higher breakeven point = strike price + (call price + put price)

Now, we briefly analyze the strategy's main features:

- *Potential profits* are theoretically unlimited because, if the underlying moves up or down beyond the breakeven point, there is no limit to the profit that can be achieved.
- *The operation risk* is limited, because maximum loss comes from the sum of put and call buying prices.
- *Potential losses* come when the underlying remains between the lowest breakeven point and the highest breakeven point range, and from a drop in implied volatility.

Let's take a practical example. When a mini–S&P 500 future is worth 1170 we buy two 1170 calls at 24.50 and simultaneously buy two 1170 puts at 24.25. Both calls and puts expire at the same time.

- Lower breakeven point: $1170 - (24.25 + 24.50) = 1121.25$
- Higher breakeven point: $1170 + (24.25 + 24.50) = 1218.75$

As you remember, to calculate the price of a mini–S&P 500 future option you need to multiply the option price by \$50. So, we will spend $(24.50 + 24.25) \times 50 \times 2 = \$4,875$, to buy two 1170 calls and two 1170 puts.

We have multiplied by two because this is the number of the option purchased in this example. Now, if the mini–S&P 500 future breaks 1218.75, which is the higher breakeven point, or falls below 11221.25, which is the lower breakeven point, our long straddle strategy will generate a profit.

When and at What Price Is It Better to Open a Long Straddle?
Books and professional traders advise you to do it when, after a lateral phase when volatility is low, it is about to face an event that everyone expects to change the low volatility, and therefore prices will rocket or plunge dramatically. These traders consider that the best timing for opening such a strategy could be just before the release of very important economic data, such as a Fed board meeting during which Mr. Bernanke could cut or hike interest rates or leave them unchanged.

Quantum Trading, instead, provides you with very powerful tools to help the trader decide the best price level to open options strategies. Quantum Trading techniques based on price, such as quantum price lines or entelechy, can show you the best price to set up a long straddle. Quantum Trading techniques based on time, as shown in Chapter 15, can help you identify algorithms and cycles indicating the end of a phase of relatively low volatility.

In any case, a sudden increase in implied volatility can further generate a profit. Assume that the mini–S&P 500 future, after the release of important economic data or news, first soars reaching 1250, and in the next two days pulls back to 1150, and then returns to 1170. These quick movements usually generate an increase of implied volatility. The change of implied volatility provides a nice surprise to the trader who had opened a long straddle because, even if the price returns to the level at which he bought the option, within a few days, by reselling the call and put options previously purchased, he can make a profit as a result of higher option prices generated by increases in implied volatility.

We could close the position and gain, due to the increase of implied volatility, even if the price moved violently within one of the two wings between the two breakeven points, without breaking them. If the market suddenly rises from 1170 to 1210 within a few hours, although it did not pass the breakeven point, implied volatility could change, which would give us a profit if we close our position even before 1218.75, the higher breakeven point, of our long straddle strategy with mini–S&P 500 options.

Adjustment of the Long Straddle Strategy   Each neutral strategy, such as the long straddle, can be adjusted. What do we mean by adjusting a strategy? At the beginning, for example, we were neutral to the expectation that the market would take a clear direction upward or downward during the rise or fall in prices. After the price breaks important support or resistance levels, we can decide to turn the strategy from neutral and nondirectional into a directional strategy.

Let's take a practical example. Suppose you buy a long straddle when the mini–S&P 500 future price is 1170. At 1219 there will be, therefore, a resistance indicated by a QPLSH. The market breaks this price and points toward the next resistance level. Using what we learned in the previous chapters about how price behaves after breaking a Quantum Price Line, we decide that it can reach the next QPL at 1260. In this case, after breaking the first resistance at 1219, we will resell the put option, and from this moment on we will generate a profit with only the bullish wing of our position. So, we close the bearish wing and we take the risk to remain with one wing open, provided that the market rises, without the hedge of the put position.

If the market falls then obviously we would have a loss from the lack of a neutral structure. But if the mini–S&P 500 rises above 1219, we are now gaining because we closed the bearish wing, and if the price reaches 1216, we gain much more by keeping only the bull wing rather than by keeping both wings.

When to Close a Long Straddle

There are two possibilities for deciding when to close a long straddle position:

- *Total closing position.* It is recommended that you completely close the long straddle following a sharp increase of implied volatility that generates a certain and immediate profit. You will sell all calls and all puts simultaneously. Then, if the price movement is very violent and has touched important supports and resistances, you may decide to close this strategy and thus simultaneously sell call and put options as the price touches a QPL or QPLSH that shows a crucial resistance or support price level. Of course, it is always convenient to do so once the price has passed above or below the higher or the lower breakeven point.

- *One wing closing position.* In case of adjustment, as we just saw, we remain with only one wing of the strategy open to maintain a strong bullish or bearish direction. Let's assume that you have opened a long straddle when the price was on a strong QPL, which could be a resistance, that is, a strategic price level from which, if broken, price would rise much higher. In this case, we are not yet sure of what could happen after the price reaches this level. We can open a long straddle just when the price touches a QPL or QPLSH and then close one wing of the strategy after the price reveals the direction it wants to follow after a few days or hours. So, if the price breaks a resistance level, we will close the put wing and keep only the long call position. If the price breaks a support, then we will close the call wing instead, keeping only the long put.

Long Strangle Strategy

This is another simple nondirectional strategy, used to open a neutral position when we set up the strategy. This allows us, under certain conditions, to gain either way, regardless if the price goes up or down.

The higher the volatility, the more you can earn. Therefore it is best to open it when the volatility is at the minimum of the period and if an increase of the implied volatility is expected.

Sometimes it's a very interesting strategy because it allows us to realize a very attractive profit with a much smaller investment compared to the long straddle. This is because purchasing OTM calls and puts is more expensive than buying ATM options.

Obviously, the risks are also greater for loss in this strategy, because the OTM options can expire OTM if their strike price is too far from the initial price.

Using QPLs, QPLSHs, and entelechy can be very useful to select the correct strike price to purchase, according to what we have explained so far.

Of course, long strangle can also be treated by adjusting the strategy after we understand that the price wants to really bounce on a very strong QPL support, or really drop after touching a powerful QPL resistance. In this case, we can close one of two wings of the strategy, liquidating the put if we see that the price continues to rise, or we can close the call position in the case that the price continues to drop.

There are no limits to the way you can use QPLs and QPLSHs to make your favorite options strategy more profitable and dramatically improve your annual return.

CHAPTER 15

A Toast to a New Achievement in Trading

This is the final chapter of the book. This is what some people might call the end of one study on how the theory of relativity and quantum physics can help you to understand the apparently erratic movements of a security. However, I hope it will be instead the beginning of a new field of research, which has already led us much further than W. D. Gann could ever have imagined. Gann left us with this brief and laconic statement, only 100 years ago:

> *Through my method . . . I can determine the price variation of every share, and taking into consideration some temporary values, I can in most cases say how a share will behave in particular conditions.*
>
> *Stocks are like electrons, atoms, and molecules that obey with continuity to their specific individuality in giving an answer to the vibration law. If we wish to be successful in our stock market speculations and avoid losses, we have to work with the causes that form the market prices. Everything in the world is based on exact mathematical proportions and everything is but a point of mathematical force.*

Many years have passed since that dinner in the spring of 2000 to celebrate a trade based on Quantum Trading theory, as described in Chapter 1. Since that evening, I have met with my friends many times, but there was a special reason that led me to invite my friends over to celebrate a new great achievement.

QUANTUM TRADING

Dave recently called me just after lunch and told me that he received my e-mail, and that he was curious and thrilled to hear about the big surprise I had mentioned. He wanted to know everything as soon as possible and he proposed that we meet for dinner—at my house, because he prefers my personal restaurant over anything else available. He loves the traditional Italian recipes that my grandmother taught me to cook, and that my personal cook has been trained to prepare to perfection.

When my friends ask me, "Why didn't you get married? You are 47; what are you waiting for?" I answer, "It's my grandmother's fault." They give me puzzled looks, wondering what I am talking about.

I respond, "My grandmother was an excellent cook. She lived with us and every day she prepared wonderful, handmade egg pasta, or fresh lasagna, ravioli, tortellini, cannelloni, risotto, and many wonderful meat dishes. When I was 10 years old, she told me, 'Now the time has come for you to learn to cook.'

"I asked her, 'Grandma, why do I have to learn to cook if you can do it so well? I don't need to.'

"She responded, 'When you grow up, you will definitely thank me. It is important to learn how to cook well, even though you are a boy. A man who is able to cook is a free man.'"

At that time, I didn't understand how right my grandmother was, because in the early 1970s in Italy, most women stayed at home and cooked for their families. It took longer in Italy than in other countries for families to adjust to the modern lifestyle in which women leave the home to take jobs in the workplace. Once it became more accepted, the American style, in which everyone is responsible for his or her own meal, spread quickly in Italy.

And so, during the summer I was trained by my grandmother to cook in the Italian tradition. Fortunately, my relatives come from different regions of Italy, because every region offers a different variety of recipes and I learned a great many. The taste, the flavor, and the ingredients change dramatically from one area to another. It's like having 20 different culinary traditions, as if there were 20 different countries in one.

When I grew up I loved cooking, food, and the finest gourmet dishes, and that's why, in my free time, I also took a sommelier's course. So, even though I had some long-term, happy relationships, I didn't get married and my girlfriends have always depended on me for cooking the best food. When I would propose going to a restaurant, they frequently preferred to stay home. After a while, I made enough money to dramatically improve my lifestyle, and I thought that a personal driver and a personal cook would be very helpful to save precious time for doing more interesting things. If you live like this, do you really want to get married? Well, you probably, like me, will postpone the moment, a moment when you will commit yourself for a lifetime, risking the transfer, if you're lucky, of at least 50 percent of your assets to the girl who officially and legally promises to love you for life.

I prefer women who promise to love me unofficially. The relationship seems to deepen and improve if you are not married. Compared with my married friends, I believe I am happier.

I have nothing against marriage, and I appreciate people who have long-term, happy relationships with their spouses. I'm positive that creating a good family and taking care of your children's education is an admirable goal that everyone should support. But in Italy and in Switzerland (where I currently live), we do not have prenuptial agreements. So, I'm waiting for the marriage laws to evolve in the direction of a civilized country, like the United States, which has already passed these enlightened laws. In the meantime, I will continue enjoying life without putting in jeopardy at least 50 percent of my personal assets. In the United States, you can lose your life savings to hucksters like Bernie Madoff, or the collapse of Bear-Stearns or of Lehman Brothers, or the collapse of the subprime real-estate–backed securities market, but believe me, these are nothing compared to not having a prenup agreement!

Coming back to Dave's phone call, I told him that having all of the crew over for dinner would be great. He asked me to prepare egg pasta with a sauce of sausage and cauliflower and I agreed. He was very curious and asked me to tell him something about my newest achievement in trading, but I refused to answer until we were all together, saying, "I will tell you everything tonight at dinner."

My friends arrived at 8:30 P.M. and I opened a few bottles of Grattamacco, a wonderful Italian red wine from the Bolgheri area in Tuscany. It is one of our favorites.

Pasta and cauliflower is very easy to prepare, and very tasty. First, steam the cauliflower. At the same time, prepare fresh, handmade egg pasta (if you are unable to do this, you can use egg noodles from the grocery store). Then, place in a large skillet extra virgin olive oil, one large piece of garlic, one-half cup of cauliflower per person, and one very tender Italian fresh pork sausage per person. Mince the sausage and saute it in the skillet together with the cauliflower, adding some paprika, pepper, and salt, and cooking it for 7 to 10 minutes on low heat. Finally, boil the pasta in a large pot and when it is finished (al dente), pour it into the skillet, cooking it together with the cauliflower and sausage sauce. Add some pepper near the end, cooking the pasta and sauce half a minute more while you mix it all together. When you serve the dish at the table, you can add some Pecorino Romano, mixed with Parmesan cheese, on top.

Elena, as usual, prepared a wonderful cake. This time she brought an *apfelstrudel*, the traditional Viennese pastry made with an apple filling.

We opened a special ice wine from Germany for desert. This is a very flavorful wine that is produced from grapes that have been frozen while still on the vine.

While Dave was still enjoying his piece of pie, Elena candidly asked, "Hey, Fabio, didn't you mention that you invited us here because you have something important to celebrate? You must tell us so that we can make a toast."

Dave, smelling the perfume of money, swallowed the rest of his pie in a single gulp and, changing his focus, immediately started to ask questions.

"Ah ha! You didn't tell us anything so far and we completely forgot why we were here because of this delicious meal. But now the meal is finished, so you have to tell us what happened! Have you found another way to make money faster and easier than QPLs and QTAs?"

"Absolutely not," I said, and watched as Dave swiftly turned his attention toward the bottle of ice wine, evidently disappointed by my news.

"So why did you invite us to toast a new trading achievement?" asked Dave.

"Well, actually what I said is not completely true. I've not discovered anything new, but I just finished the code for an automated trading system, which interfaces directly with the broker platform placing buy and sell orders, stop-loss and exit position orders, based on my Quantum Trading algorithms."

"Do you mean that you don't manually place orders on the broker's trading platform? And that your software automatically places stop-losses after the order is filled?"

"Absolutely," I answered.

"And what about intraday trading? Do you also trade futures and currencies in the very short term with your trading system, automatically?"

"Yes, sir," I responded, "I also trade in the very short term because it is important to differentiate a portfolio. Not only do I differentiate in terms of uncorrelated securities, but I can also generate a stronger alpha by simultaneously using different time frames on the same security."

"You mean that you have set up different time frames that are filtered before opening a trade on a certain security?" Dave looked impressed.

"Yes. But there is much more. In fact, I have also set up independent trading systems working with different time frames trading the same security. My automated, proprietary Quantum Trading signals range from the very short term to the medium term. So our trading time windows can range from as short as one hour to as long as six months.

"Using different time frames, we can react more quickly to changes in the trend or in volatility, both historical and implied."

"So," Dave said, "You can manage a CTA hedge fund with your automated trading system."

"The answer is yes. But it would be a different kind of CTA from the one you can find in the hedge fund industry. In fact, most CTA managers use medium and long term strategies and if the trend is choppy and doesn't develop a big movement, they lose money."

"But are there other CTA hedge fund managers using short-term strategies?"

"Very few. There aren't many CTA managers using short-term strategies for their funds. That means that when a big movement occurs, they cannot take advantage of the development of the trend, and they can lose money.

"Instead, combining the two strategies and trading the market with very short-term algorithms and medium-term algorithms at the same time, I can make money in different market conditions. Furthermore, the quantitative algos conveyed by an automated trading system isolate and generate a stronger alpha, avoiding any emotional involvement in the trade placed."

Elena requested my butler to bring some champagne from the fridge and then asked me, "Is your trading system focused more on short-term or medium-term trading? Do you prefer to follow a big trend or to trade for a very short time window?"

After many years of meeting regularly, Elena had learned quite a lot about trading, and she even started to read some books, thinking to start trading herself.

"Well, actually my trading system trades about 50 percent of the assets using medium-term algos and the remaining 50 percent is traded on very short-term and short-term time windows, such as one-minute, four-minute, and 10-minute charts. I also use 60-minute bar charts for the short term," I answered.

"The trading on this time frame is ruled by two main families of algos: contrarian algos using QPLs and TAs to forecast and trade micro-trend reversals, and trend-following algos based on QPLs breakout; that is, a signal that the trend will continue to follow the same direction with its next target being the next higher or lower QPL."

"But by increasing the frequency of the trades, can you make more money?" asked Barbara.

"It depends on how effective the algos that you rely on are. According to back-tests and many months of real trading, I can definitely say that my trading system is very profitable, based on placing various trades every day."

"And what about your trading system fitting?" Asked an ever-more interested Dave.

"Most very short-term trading systems need periodic maintenance because they are based on a bar pattern or a selection of standard indicators fitted to obtain the best result. After a while the parameters you use for the indicator need to be reviewed and you need to fit them again. Instead, with my Quantum Trading price algos and time algos, you don't need to do that because the model is quite stable and doesn't need to change its parameters every two or three months, as happens with many short-term trading systems.

"As you now know, we don't use the traditional technical-analysis tools, but we base our trading on a new philosophy, an approach inspired by the theory of relativity and quantum physics. And so, the approach is totally different. The price-deflection phenomenon and the behavior of an electron are at the base of our trading system because we consider the price to behave as if it were a photon-light particle or an electron jumping from one orbital to another."

"But enough about that," I concluded. "Now it's time to toast to my Quantum automated trading system. I wish long life, happiness, and a lot of money for everyone!"

"Cheers!"

Making Peace with Mr. Market

You're heading home after a night out at your favorite pub. To reach your car, you have to walk through a dark alley. Suddenly you sense someone approaching and shivers run through your body.

A huge, many-armed creature springs out of the shadows and accosts you! He's demanding all of your money. You're filled with fear and trembling until, suddenly, you recognize the attacker. Of course, it's the "Market!" Even as you recognize the beast for what he is, he steals your keys and runs off with your car and all of your money. The "Market" stole everything you had, but it could be worse. The real "Market" can take much more, including houses, lovers, and even wives (especially if they're young and beautiful).

The Quantum Trading techniques you are offered in this book are equivalent to a martial arts training course that will help you to defend yourself against that huge creature known as the "Market." The "Market" is a brutal monster who loves to take away money from those whom he meets on his way. But you've now learned the rules to protect yourself and how to apply them!

I'd like to ask you the following two questions:

1. Have you ever asked yourself why there are so few people in the world who are able to realize every dream they have, and live a very enjoyable, full life, both comfortable and exciting? The majority of people wake up early in the morning, come back late in the evening after a long and stressful day of work, and do this for years on end for a salary of perhaps $2,000 per month.

By contrast, privileged people can travel to every corner of the world and savor the sight of the most beautiful places on Earth, instead of living in gray and polluted cities, without worries about the cost of travel or lack of time. Everyone else lives life on a leash.

2. If you have asked yourself the first question, and you're not one of the privileged ones already, then why have you accepted this injustice until now? Why are you standing there, twiddling your thumbs instead of trying to change and become one of the privileged?

This is not an invitation to a civil riot or to a class struggle. It's an invitation to join the club reserved for those few, privileged ones who can do exactly what they want with their lives. This is because once they achieve their own "critical mass," they discover how to make money work for them, to raise their well-being.

If your answer to the second question was, "I haven't done anything to improve my life yet because I don't know where to begin," or "How could I possibly do that?" then you should stop and think about the strategies you use to enhance your experiences and to model or build your life.

The issue is not that it's impossible or even extremely difficult to be successful. People have made it before you by starting from scratch, and others will take your place if you decide not to try.

Every time I speak about money, I don't refer only to the quantitative aspect, but to the power and freedom that it can give you if you learn how to use it. Money also has the ability to improve the quality and variety of experiences that you can have in your life. This all depends on your beliefs about money, on your well-being, and on whether you believe it is possible to obtain money.

As an example, look at the Jewish people. (I want to make it clear first that I don't have Jewish ancestry.) Usually poverty occurs after being exiled from the place in which they were settled. They can also have suffered from poverty and persecution for generations. However, after moving abroad they often return to become rich. Why is this? Jewish people have very ancient roots, and they transmit to their children the belief and the conviction that wealth, well-being, and abundance are the natural status that God promised to the chosen people. For this reason they are very good businessmen.

They strongly believe that this is possible, fair, and desirable. They're consistent because they act in terms of what for them is a real and structured belief.

Whether you're religious or atheist does not matter. What matters is that you could return to Eden, the Promised Land, if only you would realize that the years of famine have ended and the exile in the desert is finished.

Who you are and what you're going to be depend only on your thoughts, on the way you perceive yourself, and on the inner image you've built for yourself. This inner image, along with your values and what you believe is possible or impossible, builds your beliefs.

If you're not part of that exclusive club of happy people who enjoy life, perhaps it's because your beliefs don't consider this as a possibility. If you really want to be part of the club, but until now you've failed, it could mean that your beliefs are defeating you. You've got to change them. To reach your goals you need time. But if you know the right strategies, then each year you can be richer, and time will be your ally.

The fear to act, to do something new, to walk in a different direction, or to start a new business is something that stops many people from even trying. Those people prefer to remain a prisoner of an illusion for years, or even for a lifetime, believing that their routine life forms a magic circle, where their safety is preserved and protected from the threats of change. For those people it is unnecessary to spend the time writing anything more. For them, everything is all right, and we respect their position.

But for those readers who hang in the balance, who have two souls, one that desires safety and another that dreams of freedom and independence, for them it is necessary to write a few more lines.

Do you often find yourself frustrated? Do you feel that the life you're living is not satisfying? Do you know that you deserve more, yet no one acknowledges your skill? Would you like to do something different, but you're terrified? Do you hesitate when it is time to make a decision to take the big step?

You think that it is almost impossible, but still you're tempted. You live, as all human beings, inside a magic circle that protects you from the demons of doubt. Inside this circle exists your entire world, created by your firm beliefs, right or wrong, that they could be made by knowledge and by experience. All that you desire is just outside of the circle, but you think that it's impossible to reach, to stretch out your hand and grab it, because the inability to leave the confines of the circle is as true for you as is the sunrise each day in the east.

The truth is that you are separated from the things you desire most by a very light and transparent barrier at the circumference of this circle. Well, maybe no one told you, or you didn't notice it, but the circle is not stiff! Its perimeter is elastic and extendable. So you can just stretch out your hand and grab what you have always dreamt. Now it can be yours.

If you simply expand the limits of your circle (which is actually nothing other than your mind, your existential models, your energy, and your emotions) suddenly, like magic, all you've ever wanted will appear inside the circle.

Do you know why? It's simply because you've widened the limits of the circle. Do you know how you can widen those limits? It's through a secret magic formula called "knowledge." The Knowledge I'm referring to is the one with the capital "K" that comes from immanent ontological thought, deep, wide, and powerful. To acknowledge means to act, to experiment, to become as the subject of your knowledge. In this way only, knowledge is power.

If you really understand what I am saying, nothing will be precluded from you, nothing will be impossible for you. You will be able to achieve your goals.

Do you know why people don't get what they want? It's because they just wish for it instead of truly wanting it. A wish is something that everyone can afford. To truly *want* something, instead, implies acting in a coherent and disciplined way, and taking chances and facing risks. Above all, whoever really wants something knows that, in any case, no matter what, he is going to get it. Every centimeter of his being and every ounce of his energy is focused on the fulfillment of his desire. This explains the difference between those who are successful and those who only dream about it.

Financial engineering at a high level is not only a method, but also an art. Throughout the chapters of this book, we have dealt with some topics that expand your knowledge of how you deal with success in trading stocks. I also hope these pages have contributed to expand the circumference of your life's possibilities. And perhaps also help you to stretch out your hand and grab what you desire most.

Good luck in trading and in life!

Acknowledgments

I would like to thank the following people who were essential to the completion of this book: Sean Boda, for helping to revise the English manuscript; Pamela Martinoli, for editing the chapter on options; and, finally, Paolo Geronazzo, for all of his support and help in preparing the charts and images throughout this book.

F.O.

About the Author

Fabio Oreste was born in Rome in 1963. He earned a Masters in Business and Finance from LUISS in Italy. Before beginning his career in finance, Fabio traveled extensively throughout the Far East, where he studied philosophy and psychology.

He now works as an international financial trader, trainer, advisor, and expert of financial engineering, and he is the founder and CEO of HF First Financial Ltd., with seats in London (UK) and Lugano (Switzerland). He has worked around the world as an independent trader and advisor of close-ended funds and structured products including notes issued by Societe Generale.

Fabio is the author of two books on trading and derivatives published in Italy: *Guadagnare in borsa in modo costante*, Rimini, Maggioli editore 2001; and *Guadagnare in borsa con le opzioni*, Milano-Il sole24Ore editore 2003.

Index

Printed in the USA
CPSIA information can be obtained
at www.ICGtesting.com
LVHW021057080224
771324LV00002B/13